PHYSICAL SCIENCE

WITH ENVIRONMENTAL APPLICATIONS

PHYSICAL SCIENCE

WITH ENVIRONMENTAL APPLICATIONS

ARTHUR W. WIGGINS *Oakland Community College* Farmington, Michigan

Houghton Mifflin Company Boston
Atlanta Dallas Geneva, Illinois Hopewell, New Jersey Palo Alto London

Printed in the U.S.A.

Library of Congress Catalog Card Number: 73-9406

ISBN: 0-395-17072-9

Photo Credits
Pages 18, 110, and 111, courtesy of the U.S. Air Force. Page 72, courtesy of Stock, Boston; photo by Daniel S. Brody. Page 261, courtesy of Professor J. W. Chapman of the University of Michigan; photo taken at the National Accelerator Laboratory, Batavia, Illinois.

Cartoons by Gahan Wilson.

CONTENTS

INTRODUCTION

To the Student

In writing this book, several assumptions have been made about you:

You are probably not going to major in the physical sciences.

Sometimes there is a negative connotation attached to being a nonmajor, a nonscientist. This book assumes nonscientists are as intelligent as scientists; they merely learn things in a different way. While scientists can study a number of separate topics in science just for the beauty of the topics themselves, nonscientists often perfer to see connections between topics, and how individual topics relate to some overall goal. Consequently, this book introduces and *connects* topics in physical science as they relate to the three principal states of matter which surround us: solids, liquids, and gases.

You have some degree of interest in environmental problems.

Man's impact on the environment is large and growing. As citizens, you are being called upon to make some decisions about our environmental policies. This book gives you a knowledge of the physical aspect of our solid, liquid, and gaseous environment to provide you with a basis for your environmental decisions. It is further hoped that you will keep this book as a reference.

Your normal reading taste runs more to novels and humorous essays than science books.

Some science books could almost be prescribed as a sure cure for insomnia. This book makes use of analogies, dialogues, socio-dramas, and bits of humor and irreverence to keep your anxiety level down and your interest up. The humor is not without purpose, though, so look for the point amidst the fun.

You wish to learn about the physical sciences to broaden your knowledge.

Although curriculum requirements may dictate this broadening, it's still a good idea. It's difficult to know what the future will bring you, and you may find some of this physical science knowledge helpful because of the added flexibility it affords.

Some of you may find this book too easy; others may find it too difficult.

Because of the wide range of skill levels people have, this book may not completely suit some people. A separate study guide has been prepared to accommodate those who find this book too easy or too difficult.

To the Instructor

This book has been developed over a period of several years at Oakland Community College. A few topics at a time were introduced in my classes, and the style and content were then modified according to student and faculty reviews.

The order in which topics are presented is nontraditional, and the exposition is intentionally a little irreverent and occasionally humorous, but not frivolous. In any event, the amount of material covered is comparable to similar texts.

I use the book for two fourteen-week terms in the following way: First term: Chapters 1–6 (solid matter); second term: Chapters 7–10 (liquids and gases) and several weeks of student-prepared presentations on topics from Chapter 10. Many students take only the first term and thus miss the liquids and gases. An alternate approach would be to give a one-semester course consisting of Chapters 1 and 2, selected sections from Chapters 3, 4, and 5, then Chapters 7, 8, and 9.

The mathematics used in this book never goes

beyond high school algebra, yet students must acquire a certain proficiency in math to understand some of the main points in the book. Mathematics is too large a part of physical science to be ignored, yet there is no point to overburdening the nonmajor student with too much math. A math refresher is built into Chapter 1 because my approach is to warn students gently, but early. If your students are better prepared, you can skip it, but a diagnostic math pre-test often reveals some real weaknesses.

I feel experiments are a central part of physical science. If you lack laboratory facilities, may I suggest that you do the experiments as demonstrations at the very least. It is important that nonscience students grasp an understanding that physical science is not only theoretical, but involved—through experiments—with real problems. Consequently, the course for which this book is designed includes a laboratory, and the experiment writeups are built right into the text.

An instructor's manual is available, giving laboratory setup instructions and equipment requirements. It also contains sample syllabi, grading systems, test questions, demonstrations, and film lists.

The study guide which goes along with this book contains supplementary material for the student who is having difficulty (a bird's eye view, worked-out problems, additional problems, and self-tests) and challenging material for the student who finds the material too easy (mind-stretching questions, tangential but interesting topics, project suggestions, and other sources to dig into).

Arthur W. Wiggins

ACKNOWLEDGMENTS

Thanks are due to many people, without whose help this book could never have been done: Karen Esshaki for typing; "Doc" Barry and Jim Schulz for cartooning in preliminary editions; Charlie Wynn, Rich Wells, and Al Varone for general support and encouragement; Teri Davis for the index; and my family, Reggie, Ann, and John, for their patience. Last but not least, thanks to my colleagues, friends, and students who made so many good suggestions along the way.

For their help as reviewers at various stages of the manuscript, I would like to thank Bill G. Aldridge, Florissant Valley Community College, Missouri; Merrill Chamberlain, Ball State University, Indiana; Norris W. Goldsmith, State University of New York at Oswego; Vallie W. Guthrie, North Carolina Agricultural and Technical State University; Marvin L. Ivey, St. Petersburg Junior College, Florida; Rod O'Connor, Washington State University; Marvin Paul, The Loop College, Illinois; R. J. Ruch, Kent State University, Ohio; and Peter Zito, Nassau Community College, New York.

CHAPTER ONE

WHAT IS PHYSICAL SCIENCE ALL ABOUT?

Man has always had a strong curiosity about what is going on around him. Since man has been studying things around him for thousands of years, rather large bodies of knowledge and information have been built up regarding man's surroundings. To bring some kind of order to all this information, several large divisions have been made, as shown in Table 1.1.

Social sciences study the nonmaterial or intellectual part of man's environment, while biological sciences study the living part of man's material

Table 1.1 *Relationships Among Subdividions of Man's Environment*

			STUDIED BY
MAN'S ENVIRONMENT	Nonmaterial		Social sciences (history, sociology, economics, political science, psychology, etc.)
	Material	Living	Biological sciences
		Nonliving	Physical sciences

environment and physical sciences study the nonliving, material part of man's environment. This subdivision is rather broad, and there is a great deal of overlap, but these difficulties only serve to emphasize the fact that these are *large* bodies of knowledge we're dealing with.

The goal of the entire book is to start you thinking about the material, nonliving part of our environment along the lines that physical scientists think. This first chapter will lay the groundwork by looking at the way physical science has evolved its intellectual methods, particularly through the scientific revolution. We'll then look at the "scientific method" as it is used in modern times; we'll see why some mathematics is necessary to physical science, and the importance of systems of units. Last, we'll do an experiment to illustrate how the "scientific method," mathematics, and units can combine to solve a practical problem.

The aim of this chapter is to accomplish some necessary preliminaries and get us started thinking in the way we'll have to think to accomplish our overall goal of analyzing solid, liquid, and gaseous matter and applying our knowledge to environmental problems.

1.1

In the Beginning . . .

Our earliest ancestors required some knowledge about the physical portion of their environment so that they could figure out when to plant crops, what stones would make arrowheads, and which liquids freeze in winter. Today, some feel that we don't need to know very much, as illustrated by C. H. Ump* in Figure 1.1.

C. H. Ump might regret his ignorance if the smog cloud envelops him, or if the open dump causes the nearby river to become polluted. The point of this illustration is that we still need some knowledge about our physical environment even in our modern technological society, and perhaps we need this knowledge more than ever to assess our own impact on our environment.

*Note: A whole family of characters whose last name is Ump will appear periodically in this book to take a slightly whimsical view of some of the concepts presented here. Please study these illustrations carefully, as their message is sometimes subtle, but almost always relevant.

Figure 1.1 *C. H. Ump comments on his environment*

Although the earliest studies of the physical environment were very practical, as man's sophistication grew, some people didn't have to work at the ordinary tasks of hunting, farming, and soldiering, and so were able to think about the larger, more philosophical issues.

The portion of philosophy which dealt with the environment was called *natural philosophy*. Although some natural philosophers did some experiments, their theories and arguments were done chiefly on an intellectual plane, particularly in the West. Their ideas were presented and discussed in the light of the writings of previous natural philosophers. A prime example of a natural philosopher who did no experiments but theorized on the basis of previous writings was Aristotle† (about 350 B.C.), who propounded the belief that two bodies falling freely would fall in such a way that the heavier body would fall faster than the lighter one, as shown in Figure 1.2. Aristotle's theory was based on intellectual arguments rather than extensive experimentation.

Figure 1.2 *Aristotle theorizing that a heavy body falls faster than a light one*

Because of the necessity of agreement with previous natural philosophers, early physical science changed very slowly.

1.2

Scientific Revolution, 1600 Style— "Scientific Method"

Western physical science underwent a dramatic change in the 1500's and 1600's. The dramatic change was in the intellectual method used by scientists. Just as other parts of the Western civilization were changing, moving away from an authority-dominated structure, physical science moved away from the authoritarian natural philosophy to a method whose final test of truth is *experiment,* not the writings of a previous philosopher.

This revolutionary change in science's intellectual method

†Note: Short biographies of people will *not* be given in this book because they are available in any encyclopedia or in some good dictionaries, such as the American Heritage Dictionary, and should be looked up if the student is unfamiliar with their names.

was not accomplished by one single individual, nor was it achieved in one single year or decade. The point is that the tinkerers and the practical men who worked with various kinds of equipment had been considered "crackpots" before the revolution. After the revolution, experimentation was not only respectable, it was necessary to establish the validity of scientific theories. As an example of the importance of experimentation *after* the revolution, let's look at Galileo's treatment of the problem of freely falling bodies.

Whereas Aristotle *theorized* (or at best conducted limited experiments) that heavy bodies fall faster than light ones, Galileo began dropping bodies of various mass and measuring their motion. Some historians believe that Galileo was theatrical enough to drop some bodies from the Leaning Tower of Pisa, as shown in Figure 1.3.

Figure 1.3 *Galileo drops bodies of different mass from the Leaning Tower of Pisa*

Galileo's result was that bodies of different mass fall at the same rate. This finding differed from the theory of Aristotle. More important than the speed of falling bodies is the fact that Galileo used an entirely different method to achieve his results—*experimentation.*

The method used by physical science since the first scientific revolution is often called the *"scientific method."* There are some unfortunate and inaccurate connotations to the term "scientific method" which we must clarify before we go any further. The scientific method does *not* operate like a computer, in which problems are fed in one end and solutions appear automatically, almost magically, at the other end. It does work much like the machine shown in Figure 1.4.

One way of characterizing the scientific method is in terms of a recycling sequence of four activities:

Observation
Hypothesizing
Prediction ⟲ Recycle
Experimentation

Observation is somehow sensing (seeing, hearing, feeling, recording, etc.) a natural occurrence.

Hypothesizing is the construction of a conceptual model of the basic causes of the phenomenon observed. The hypothesis should be as general as possible to explain other phenomena besides the specific one observed.

Prediction is the application of the hypothesis to a phenomenon other than the one which gave rise to the hypothesis in the first place.

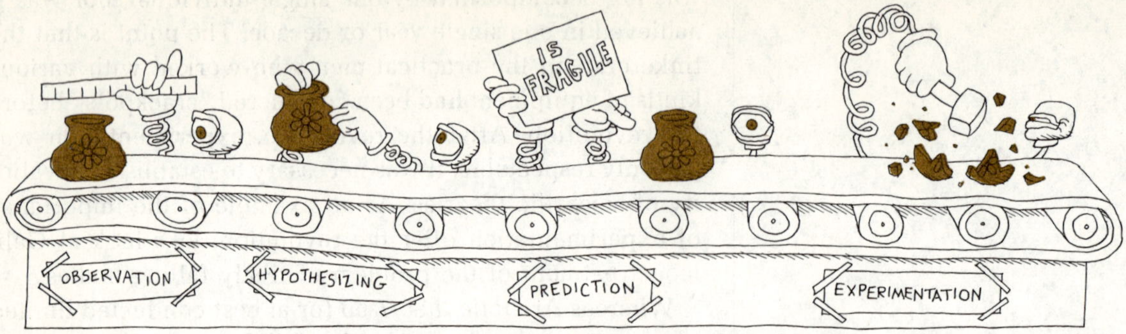

Figure 1.4 *Scientific method machine*

Experimentation is the test devised to see if the prediction is true or false by looking for the occurrence of the phenomenon predicted above.

Recycling occurs if the experiments yield results which differ from the predictions. The hypotheses must be modified to explain the experimental results. Once the hypothesis is modified, new predictions may be made and new experiments performed to check out the predictions.

Before we see all the little modifications to this sequence, let's look at a simple example of how the scientific method might work in some sort of ideal way.

Suppose Newton sat under an apple tree and *observed* that an apple fell and hit him on the head. He then *hypothesized* that the apple and the earth attracted each other, and that all bodies attract each other. He could then have *predicted* that a small mass attached to a pointer would be attracted to a large boulder, causing the pointer to move. If he had performed that *experiment*, he would have found that the small mass would be attracted by the boulder, causing the pointer to move, thus verifying his prediction and hence his hypothesis.

The main difficulty with the nice, neat, little sequence we have been looking at is that scientists don't work in such an orderly fashion. It is often possible to go back and identify each item of the scientific method sequence in the development of a particular theory; however, in many cases the items are accomplished by different people, in different places, at different times, out of sequence, and sometimes the people were unaware of the existence of the others' work. For example, the theory of relativity could be thought of as beginning with an experimental observation made in 1887 by Michelson and Morley, followed by Einstein's hypothesis in 1905, and partial verification from astronomical data on the orbit of the planet Mercury. Yet, Einstein has been quoted to say that he was not particularly influenced by the Michelson-Morley experiment, and the astronomical data wasn't measured by Einstein.

Another difficulty with the scientific method lies in the relationship between scientific truth and *certainty*. We are all familiar with the type of reasoning typified by the example: "All dollar bills are green. That is a dollar bill; therefore, that dollar bill is green." This is called *deductive reasoning;* it begins with a general statement (all dollar bills are green) and draws a specific conclusion (that dollar bill is green). Note that deductive reasoning proceeds from the general to the specific, and there is no uncertainty involved, and no new information is generated.

On the other hand, *inductive reasoning* proceeds from a certain specific truth to an uncertain generality. For example: "I have seen 99 mice, and they have all been gray. Therefore, all mice are gray." The conclusion is general, but not certain. Suppose mouse number 100 scurries past our observer, and it is white. Our observer must modify his hypothesis: "I have seen 100 mice, 99 gray and 1 white." Therefore, 99% of all mice are gray and 1% of all mice are white.

The scientific method is an example of inductive reasoning, because the general hypotheses are advanced on the basis of specific observations, and the hypotheses are, therefore, only as good as the observations. In a sense, the reason experiments are done is to broaden the base of experience from which the hypotheses come, and to provide opportunity for modification of the hypotheses to reflect reality more accurately.

We can see now why the scientific method is *not* some kind of magical machine able to accept problems and crank out solutions. The hypothesis doesn't just jump out of the observations; it must be created, sometimes by the stroke of a genius. The creation of a good hypothesis by a scientist is in many ways comparable to a composer getting the notes just "right" for his new song or a painter getting the shading just "so" in his portrait. Hypothesis-making is creative and hence cannot be scheduled or programmed like a machine.

It is also possible to see why the scientific method has not been applied to the social sciences with the same degree of success as it has enjoyed with the physical sciences. In the physical sciences, hypotheses may be tested by running laboratory experiments under carefully controlled conditions, varying only one quantity at a time. In many of the social sciences (e.g., economics, psychology) there is no equivalent laboratory in which to test hypotheses because humans are involved; and we are a mixture of so many quantities, that it is hard to hold all but one quantity constant.

Because the scientific method is so deeply concerned with physical *quantities*—observing them, hypothesizing relationships between them, predicting their values, measuring them in experiments—we're going to take a closer look at two aspects of quantities that may be a little unfamiliar to us or a little rusty in our memories: *mathematics,* shorthand ways of hypothesizing relationships between quantities; and *systems of units,* standard ways of measuring quantities so that relationships may be confirmed or rejected by any experimenter on a consistent basis, and so that experimental results may be compared easily.

1.3

Mathematics Refresher: A Dialogue

EAGER STUDENT: Well, I guess we might as well get on with it. What kind of math will this book use?

PROFESSOR: Just simple algebra. All equations will involve simple sums or unknowns to the first (x) or second power (x^2).

STUDENT: That's good; equations mess up my mind.

PROFESSOR: They won't if you think of equations as if they were a balance. If you do something to one side only, you no longer have a balance.

In Figure 1.5, the left side was divided by 3, the right side left alone; so it no longer balances. If, instead, both sides had been divided by 3, it would remain balanced. The point is that if you divide one side, you must divide the other; if you multiply one side, you must multiply the other, etc., in order to retain the balance, or else you have no equation. In other words, *do the same thing to both sides of an equation.*

STUDENT: That sounds easy enough, but how did you know it should be divided by 3 to get x all alone?

PROFESSOR: This is called *isolation of the unknown,* and please bear with a corny story which illustrates this point:

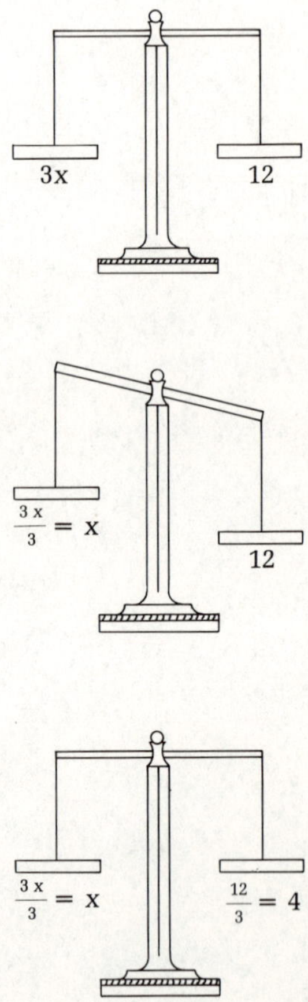

Figure 1.5 *Equations as physical balances*

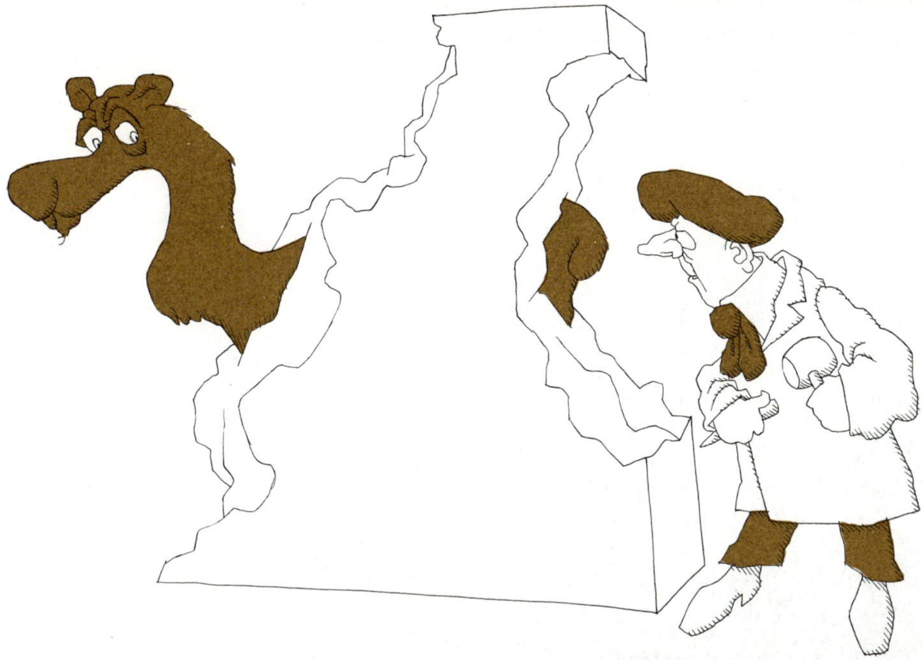

Figure 1.6 *Sculpture of H. Ump*

A famous sculptor sculpted a statue of the camel H. Ump from a block of marble (Figure 1.6). The critics hailed it as a major work of art and asked him how he went about it without any model. "It was easy," he said; "I just chipped away anything that didn't remind me of H. Ump." (Pause until groans subside.)

The significance of this story is that, as the sculptor isolated the camel by chipping everything else away, the unknown is isolated by chipping away all the knowns.

Example #1:

$$17z = 34.$$

The z is the unknown, so we have to chip away the 17 to get at it. Since 17 multiplies z, we must divide by it to chip it away, and we have to do it to both sides to retain our balance.

$$\frac{17z}{17} = \frac{34}{17},$$

$$z = 2.$$

Example #2:

$$\frac{\alpha}{4} = 62.$$

The α is unknown, so we have to get rid of the 4. Since it divides the α, we'll multiply by it (both sides, you remember).

$$\frac{4\alpha}{4} = 62(4),$$

$$\alpha = 248.$$

Example #3:

$$w + \frac{1}{2} = 2.$$

Since $\frac{1}{2}$ is added, we must subtract it to isolate the w.

$$w + \frac{1}{2} - \frac{1}{2} = 2 - \frac{1}{2},$$

$$w = 1\frac{1}{2}.$$

Example #4:

$$a - 6 = 0.007;$$

adding 6, $$a - 6 + 6 = 0.007 + 6,$$

$$a = 6.007.$$

Example #5:

$$3y + 8 = 17.$$

We have to get rid of both the 3 and the 8, and it's usually easier to get rid of the additive or subtractive term first.

$$3y + 8 - 8 = 17 - 8,$$

$$3y = 9,$$

$$\frac{3y}{3} = \frac{9}{3},$$

$$y = 3.$$

EAGER STUDENT: It seems easy enough with those good numbers, but why do you use all the different symbols?

PROFESSOR: Because we're about to pull two fast ones on you.

Fast One #1: About those symbols . . . Elementary algebra often has you solve equations where the unknown is represented by a letter (sometimes a Greek letter for variety), and the knowns all have numerical values. In physical science, equations use mostly letters which represent physical quantities, and the unknown may vary from one problem to the next.

Example #1:

$$D = \frac{m}{V}.$$

If D and V are given and hence known, you must solve for m by chipping away the V.

$$DV = \frac{m}{V} \cdot V,$$

or,

$$DV = m.$$

If $D = 8$, $V = 6$, then $m = DV = 8(6) = 48$. Or, if D and m are given, you must solve for V.

$$DV = m,$$

$$\frac{DV}{D} = \frac{m}{D},$$

$$V = \frac{m}{D}.$$

So, if $m = 48$, $D = 8$, $V = {}^{48}\!/_{8} = 6$.

Example #2:

$$PE + KE = TE.$$

Given

$$KE = 16,\ TE = 28,$$

then

$$PE + KE - KE = TE - KE,$$
$$PE = TE - KE = 28 - 16 = 12.$$

Or, if

$$PE = 12,\ TE = 28,$$

then

$$PE + KE - PE = TE - PE,$$
$$KE = TE - PE = 28 - 12 = 16.$$

So, in general, we'll deal with equations using symbols rather than numbers, because we won't always have the same unknown.

Fast One #2: Those simple numbers just can't last. In this book, we'll deal with some very large numbers, for example: 602,000,000,000,000,000,000,000; and some very small numbers, for example: 0.0000000003.

Few people enjoy writing all those zeros, and, furthermore, some may get lost, leading to wrong results. So, some clever scientist invented a way around writing out all those zeros, and this way is called the *Standard Scientific Notation.* It consists of writing numbers in the form:

$$6.7 \times 10^2.$$

6.7 is called the *coefficient,* and must be between 1 and 9.99999.

10 is the *base.*

2 is the *power* or *exponent.*

This is particularly useful for large or small numbers or multiplying and dividing. For example:

$$602{,}000{,}000{,}000{,}000{,}000{,}000{,}000 = 6.02 \times 10^{23}.$$

To get the exponent, simply count how far you have to move the decimal to get the coefficient the right size. Another example:

$$0.000038 = 3.8 \times 10^{-5}.$$

You can think of this as multiplying and dividing by the same number, which leaves the original number unchanged.

$$\frac{0.000038}{100{,}000} \times 100{,}000 = \frac{3.8}{100{,}000} = 3.8 \times 10^{-5}.$$

Multiplying or dividing two or more numbers becomes particularly simple when you use standard scientific notation.

Multiplication

Procedure

1. Multiply coefficients
2. Carry along base
3. Add exponents (watch the signs!)

Example:

$$(6 \times 10^{23})(4 \times 10^{-4}) = 24 \times 10^{23-4},$$

$$= 24 \times 10^{19},$$

$$= \frac{24}{10} \times 10^{19} \cdot 10^{1},$$

$$= 2.4 \times 10^{20}.$$

Division

Procedure

1. Divide coefficients
2. Carry along base
3. Subtract lower exponent (*denominator*) from upper (*numerator*) exponent (watch the signs!)

Example #1:

$$\frac{2.12 \times 10^{8}}{1.06 \times 10^{-8}} = \frac{2.12}{1.06} \times 10^{8-(-8)} = 2 \times 10^{16}.$$

Example #2:

$$\frac{8.4 \times 10^{-16}}{2.1 \times 10^{-14}} = 4 \times 10^{-16-(-14)} = 4 \times 10^{-2}.$$

Example #3:

$$\frac{6.3 \times 10^{12}}{9 \times 10^{4}} = 0.7 \times 10^{8} = 7 \times 10^{7}.$$

EAGER STUDENT: It does look easier to use standard scientific notation for either very large or very small numbers.

PROFESSOR: Good, because we'll see plenty of both before the book is over.

STUDENT: Whoopee, I can hardly wait! Do you have any more mathematical tricks to pass along?

PROFESSOR: Yes, we need to discuss proportionality and graphs.

If two quantities are related in such a way that one quantity increases as the other quantity increases, then the two quantities are said to be directly proportional. For example, as your income increases, your tax increases; right?

STUDENT: That's true; you certainly picked a clear example.

PROFESSOR: Sorry; nevertheless, since your tax increases as your income increases, it could be said that your tax is directly proportional to your income.

(tax) directly proportional to (income)

Naturally enough, there is a shorthand symbol for "directly proportional to." It is $\propto$.

$$(\text{tax}) \propto (\text{income})$$

STUDENT: I think I understand that. Directly proportional means that two quantities increase or decrease together.

What were you going to say about graphs?

PROFESSOR: Graphs present a lot of experimental data in a form that is neat, concise, and very illustrative of developing patterns. For example, let's look at some (imaginary) data on the rainfall on Umpville, U.S.A.

YEAR	RAINFALL (inches)	YEAR	RAINFALL (inches)
1964	8.6	1969	31.0
1965	16.4	1970	29.2
1966	12.7	1971	36.7
1967	20.3	1972	41.2
1968	24.7	1973	43.7

To make a graph of this, we'll need some graph paper, then we'll have to choose a uniform scale in the horizontal direction for the years.

STUDENT: There are lots of lines on this paper; how many lines should I choose for one year?

PROFESSOR: The number of lines that you choose to represent one year will determine the horizontal size of the graph. Try a couple of different choices and see what happens.

Now choose a uniform scale in the vertical direction for the inches of rainfall. Again, the choice of the number of lines that represent one inch determines the vertical scale of your graph.

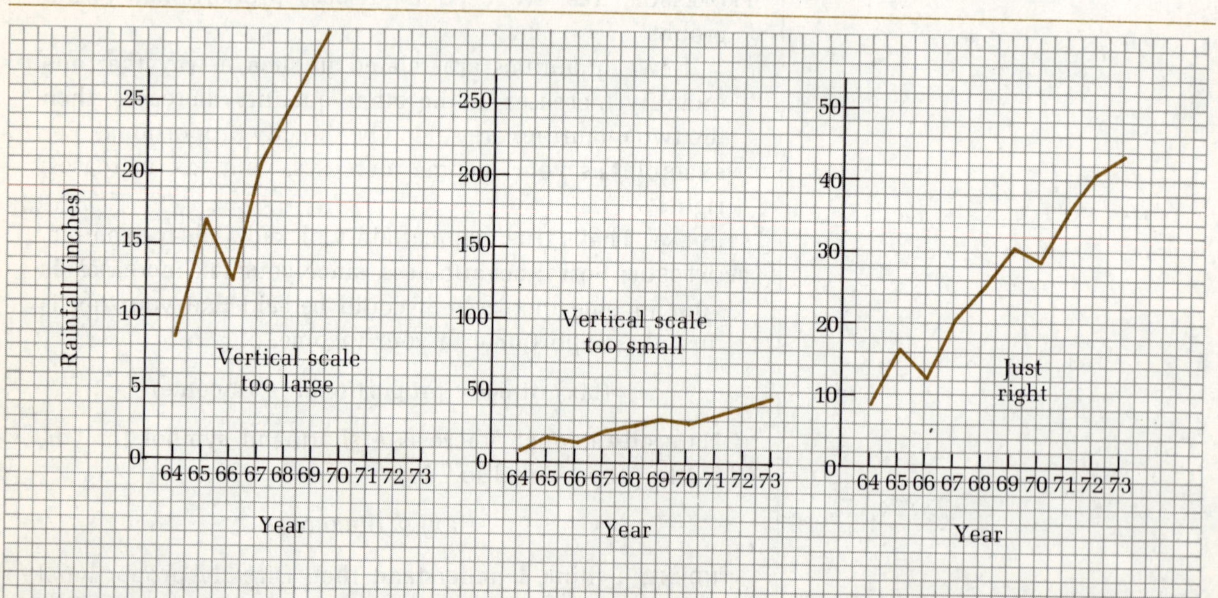

Figure 1.7 *Graphing*

STUDENT: Oh, I see. If you choose the years and inches too close together, you wind up with a graph all cramped down in a corner; but, if you choose them too far apart, you could go off the paper.

PROFESSOR: In Figure 1.7, I've illustrated this graph and some common mistakes in graphing.

By the way, do you see any pattern emerging from this data?

STUDENT: Well, if I lived in Umpville, I think I'd start building an ark.

1.4

Measurement of Quantities: The Metric System

Suppose you asked someone how far it is from his home to the airport, as shown in Figure 1.8, and he answered "6." You would be understandably confused, because he could mean 6 miles, 6 minutes, 6 hours, or even 6 meters.

The point is that there are two necessary parts to any

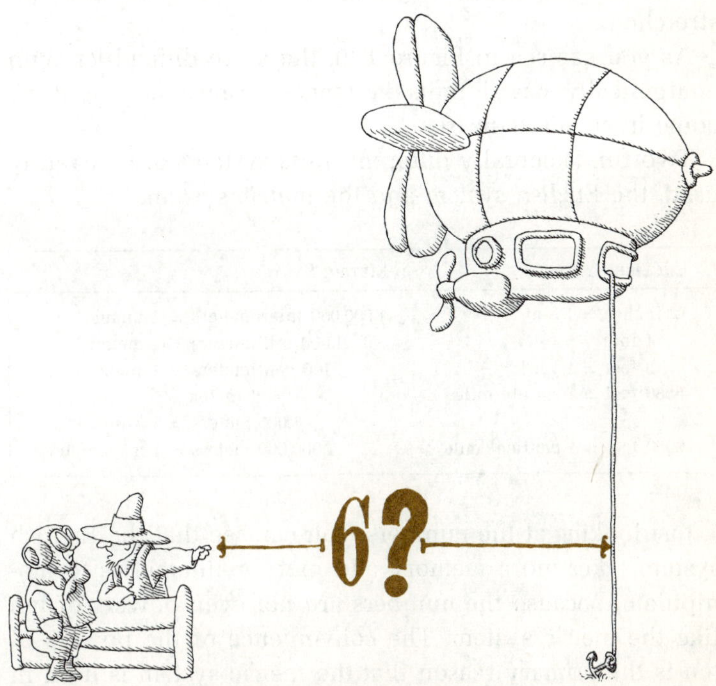

Figure 1.8 *Six what?*

measured quantity, a numerical part and a units part. For example:

Numerical	Units
6	Minutes
14	Light-years
1600	Dollars

The units part of quantities must balance on both sides of an equation, as well as the numerical part. Keeping track of units, called *dimensional analysis,* is a handy way to make sure equations are working out correctly.

In the physical sciences, the three quantities chosen to be fundamental are *length, mass,* and *time.* Any system of units must measure *all* quantities in terms of the units assigned to measure these fundamental quantities.

Length

Historically, standard units of length were based on various anatomical distances as shown in Figure 1.9. An inch was the distance from the tip of the thumb to the first knuckle. A foot was the length of a British king's foot. A cubit was the distance from the elbow to the fingertip. A fathom was the distance between fingertips when the arms are outstretched.

Figure 1.9 *Some units*

As you can see in Figure 1.10, there are difficulties with anatomically based units systems, because people don't come in standard sizes!

Two fundamentally different units systems are currently used, the *English system* and the *metric system.*

English System	Metric System
12 inches = 1 foot	1,000,000 micrometers = 1 meter
1 foot	1,000 millimeters = 1 meter
3 feet = 1 yard	100 centimeters = 1 meter
5280 feet = 1 statute mile	1 meter
	1,000 meters = 1 kilometer
6080 feet = 1 nautical mile	1,000,000 meters = 1 megameter

Figure 1.10 *One mini-fathom*

Just looking at the numbers, you can see that the English system takes more memory and more mathematics to manipulate, because the numbers are not even powers of ten like the metric system. The convenience of the powers of ten is the primary reason that the metric system is used in

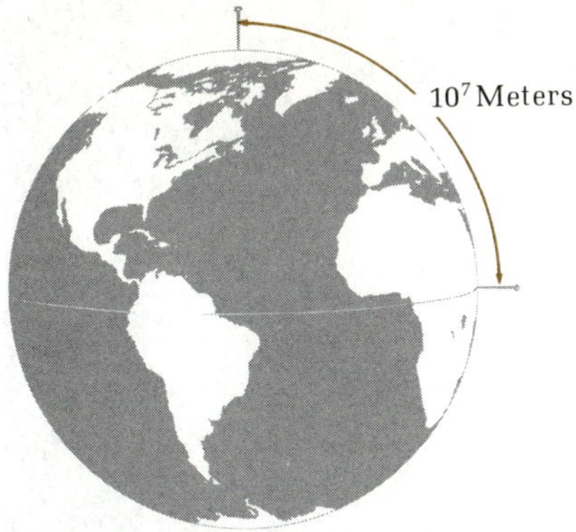

Figure 1.11 *Meter definition*

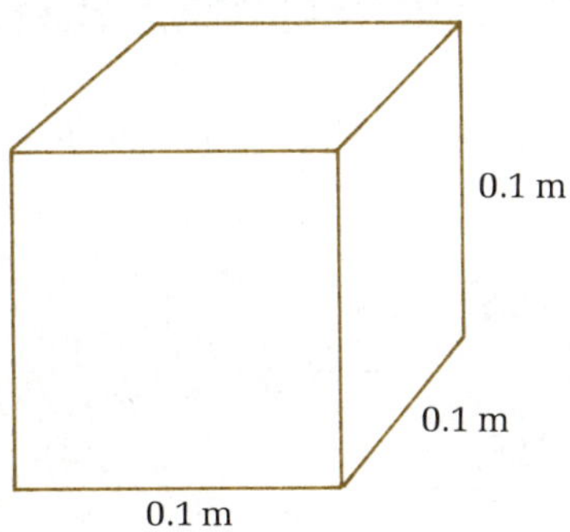

Figure 1.12 *Metric conversion*

science. (Scientists are as lazy as other people and don't like to get "hung up" in all sorts of crazy conversion factors.)

The meter was originally defined just before 1800 by the French Academy of Science as a ten-millionth of the distance from the pole to the equator as shown in Figure 1.11. On a more practical level, they made two scratches on a special metal bar, and the distance between the scratches was defined as one meter.

Although we won't convert back and forth, a meter corresponds to just over 3 feet (about 39.4 inches) in the English system. (See Figure 1.12 about this conversion.)

Mass

The English system actually measures weight rather than mass, but the two are directly proportional to each other. (The relationship between weight and mass will be further explored in Chapter 4.)

English System (weight)	Metric System (mass)
16 ounces = 1 pound	1,000,000 micrograms = 1 gram
1 pound	1000 milligrams = 1 gram
2000 pounds = 1 ton	100 centigrams = 1 gram
	1000 grams = 1 kilogram
	1,000,000 grams = 1 megagram

The basic unit of mass defined by the French was the *kilogram*. A kilogram is the mass of water contained in a cube $\frac{1}{10}$ meter on a side. Again, the metric system has a completely decimal conversion. In fact, the decimals are simply indicated by the prefix in front of the unit.

Micro = 1 millionth = 10^{-6}
Milli = 1 thousandth = 10^{-3}
Centi = 1 hundredth = 10^{-2}
Kilo = 1 thousand = 10^{3}
Mega = 1 million = 10^{6}

These prefixes are so handy that we tend to use them to describe other things besides metric measurements. For example, the atom bomb's destructive capability is measured in *kilo*tons of TNT (see Figure 1.13); H-bombs are measured in *mega*tons of TNT. Girls in very short skirts are said to be wearing *micro*skirts (probably an exaggeration).

An easy way to convert measurements from one prefix to another is to simply relate the desired prefixes to some

Figure 1.13 *Atom bomb test*

number of the fundamental units, then eliminate the fundamental unit. For example:

How many micrometers are there in one millimeter?

$$10^6 \text{ micrometers} = 1 \text{ meter,}$$
$$\text{and } 10^3 \text{ millimeters} = 1 \text{ meter,}$$
$$\text{therefore } 10^6 \text{ micrometers} = 10^3 \text{ millimeters.}$$
$$\text{Dividing by } 10^3,\ 10^3 \text{ micrometers} = 1 \text{ millimeter.}$$

or

How many milligrams are there in a kilogram?

$$10^3 \text{ milligrams} = 1 \text{ gram,}$$
$$10^3 \text{ grams} = 1 \text{ kilogram.}$$
$$\text{Multiplying by } 10^3,\ 10^6 \text{ milligrams} = 10^3 \text{ grams;}$$
$$\text{so, } 10^6 \text{ milligrams} = 1 \text{ kilogram.}$$

Notice that the conversions involve powers of ten *only*, without any messy coefficients.

Time

The most natural unit of time is the day, but decimal subdivisions and multiples of a day (e.g., a milliday is approximately equal to a minute and a half; a kiloday is approximately equal to almost 3 years) have never really caught on. Traditionally, we use the *second*, 1/86,400 of a day, in both the English and metric systems.

The metric system is used so extensively in the world today that the United States will probably be following this system by the 1980's. The space program has already adopted the metric system, and Congress has enacted legislation to gradually change to the metric system. After all, the English, who originated the English system, expect to be fully converted to the metric system in the near future.

1.5

Density is Our Destiny

Now that we have polished our mathematical skills and straightened out a few things about units systems, let's apply these tools to a quantity that will provide us with some practice now and will be useful to us later—*density*.

Suppose a lumber baron bets you $100 that you can't determine the mass of a large log by using only a small scale, a meter stick, and a small saw. Take the bet! Within the next few pages, you'll find out how to collect.

Envision a cube of Jello (your favorite flavor) shimmering happily at you from its place in your bowl (Figure 1.14). This chunk of matter has some important properties. First, it has a certain amount of "stuff" or matter to it; this is called its *mass*. Second, it takes up a certain amount of space; this is called its *volume*. The ratio of the mass of the Jello cube to its volume is called *density*.

$$\text{Density} = \frac{\text{Mass}}{\text{Volume}} = \frac{\text{Amount of matter}}{\text{Space occupied}}$$

Now, if you had two cubes of Jello, the mass would double and so would the volume, so the density would be multiplied by $\frac{2}{2}$ and, therefore, unchanged. In fact, if you had three cubes, or any other number, the mass and volume would continue to rise proportionally, and their ratio would remain constant. Thus, we are led to the conclusion that *density is a property of the material, independent of the amount of material present.* (Aha, the lumber baron's $100 is in jeopardy! We can cut a small chunk out of the big log, and the chunk will have the same density as that of the log.) In terms of the shorthand used by scientists, $D = m/V$. Since quantities must have units, the units of density must be kg/m^3, because mass is measured in kilograms and volume in m^3. Note that any one of the three quantities may be solved for by manipulating the equation. For example:

Figure 1.14 *Jello cubes in a bowl*

Given D and V, solve for m. Multiply both sides by V:

$$DV = \frac{m}{V} \cdot V,$$

or

$$m = DV.$$

Given m and D, solve for V. Divide both sides by D:

$$\frac{m}{D} = \frac{DV}{D} = V,$$

or

$$V = \frac{m}{D}.$$

Example #1: Given a 2 kg beef roast whose volume is 2×10^{-3} m³, what is the roast's density?

$$D = \frac{m}{V} = \frac{2 \text{ kg}}{2 \times 10^{-3} \text{ m}^3} = 1 \times 10^3 \frac{\text{kg}}{\text{m}^3}.$$

Example #2: A 10 m³ boulder has a density of 3000 kg/m³. What is the boulder's mass?

$$m = DV = 3000 \frac{\text{kg}}{\text{m}^3} \cdot 10 \text{ m}^3 = 3 \times 10^4 \text{ kg}.$$

Example #3: Air has a density of about 1 kg/m³. What volume of air would it take to get a mass of 100 kg?

$$V = \frac{m}{D} = \frac{100 \text{ kg}}{1 \frac{\text{kg}}{\text{m}^3}} = 100 \text{ m}^3.$$

Once you understand two little tricks, you'll get your shot at the lumber baron's $100:

Trick #1: No volume given, just dimensions of some object. It's easy enough to look up the volume of regular solids. See Figure 1.15 for some easy ones.

Example: A metal cylinder 20 cm in diameter and 50 cm long has a density of 3 g/cm³. What is the mass of this cylinder?

$$m = DV = 3 \text{ g/cm}^3 \cdot \pi(10^2) \text{ cm}^2(50 \text{ cm}) = 4.71 \times 10^4 \text{ g}.$$

Trick #2: Mismatched dimensions. Although the fundamental units for density are kg/m³, the units often used are g/cm³. The reason for this is that water has a density of

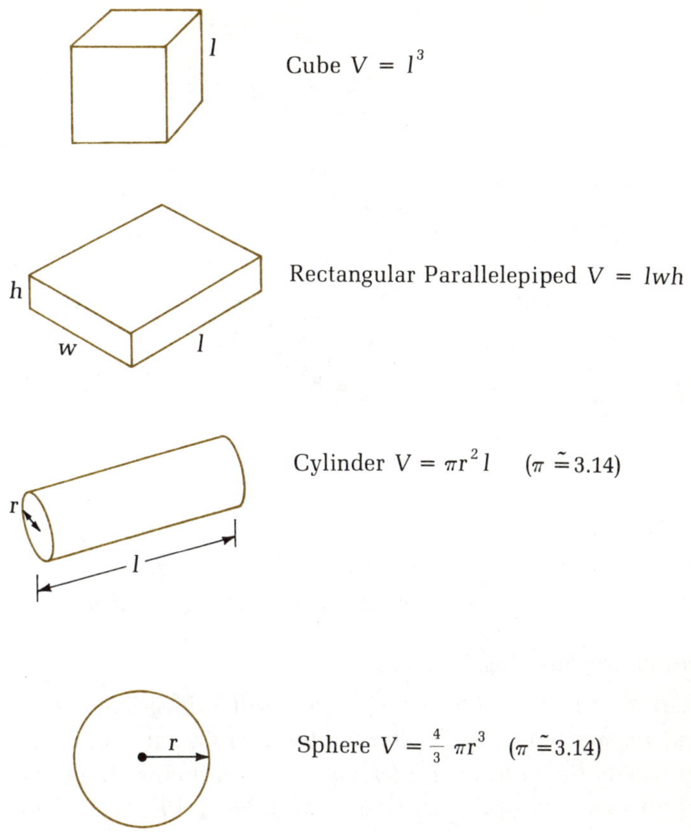

Figure 1.15 *Volumes of some common solids*

1 g/cm^3 (you can show that this corresponds to 1000 kg/m^3) and water is a convenient reference.

If some dimensions are stated in meters, others in centimeters, masses in kg, densities in g/cm^3, etc., all you really need to do is to convert the units until they are all compatible and proceed with the solution.

Example: A rectangular chunk of polystyrene plastic is 10 cm $\times$ 40 cm $\times$ 2 m. If its density is 2×10^{-2} g/cm^3, what is its mass in kg?

$$m = DV = 2 \times 10^{-2}\ \text{g/cm}^3\,(10\ \text{cm})(40\ \text{cm})(200\ \text{cm})$$
$$= 1600\ \text{g} = 1.6\ \text{kg}.$$

EXPERIMENT 1.1

THE LUMBER BARON'S $100 LOG

With your handy-dandy meter stick, suppose you measure the lumber baron's log (Figure 1.16) and find that it is a cylinder 80 cm in diameter and 20 m long. Using a balance and meter stick on the piece of wood sawed out of the log

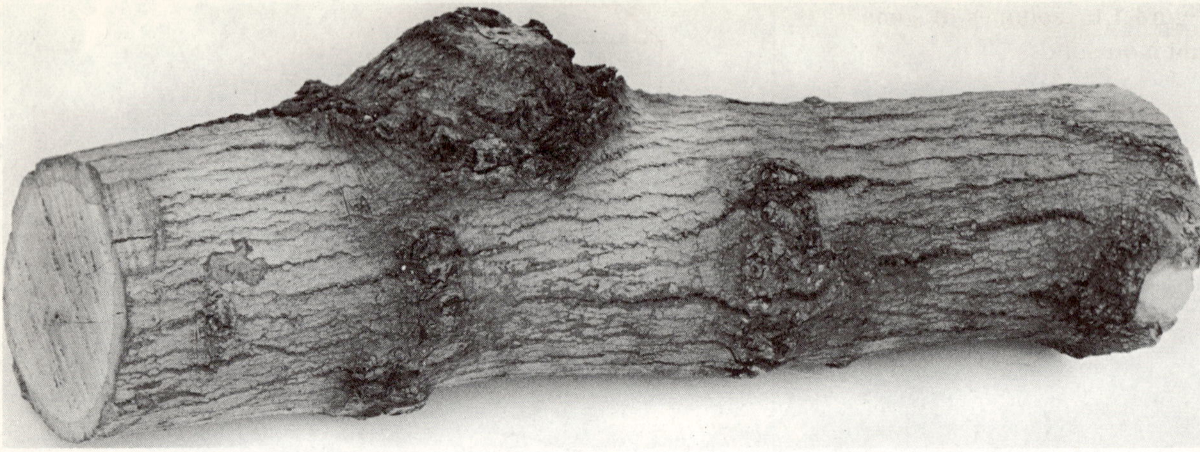

Figure 1.16 *The lumber baron's $100 log*

(to be supplied by the instructor), use the methods developed in this section to find the mass of the large log, and to win the $100 from the lumber baron—if you can find him.

Double or Nothing

Your instructor will also supply you with a large rock which was formerly part of a spherical boulder 4 meters in diameter. Using the balance and some additional equipment, calculate the mass of the original boulder. (Hint: The liquid measure unit 1 milliliter (ml) = 1 cm^3*.)*

Summary

Large bodies of information about man and his environment have been built up over the years. This information has been divided into social sciences, which study the nonmaterial parts of man and his environment; biological sciences, which study the living part of man and his material environment; and physical sciences, which study the nonliving parts of man's environment.

Early man required some practical knowledge of his physical environment for purely practical reasons, but some feel that modern man doesn't need to know as much. Many others think that our modern technological society is having such a large effect on our environment that we need more knowledge of physical science than ever before.

As man became more civilized, some men took time to think about larger philosophical issues, and natural philosophy was started. Early natural philosophers worked almost exclusively on an intellectual plane. Aristotle advanced the theory that heavy bodies fall faster than light ones, but performed no systematic experiments on falling bodies.

In the 1500's and 1600's, Western physical science underwent a revolutionary change in its method—experimentation was recognized as the final test of the truth of a theory. Galileo experimented with falling bodies and found that they fell at the same rate regardless of their mass.

The "scientific method," developed through the experimental revolution, refers to a recycling sequence of activities: observation, hypothesizing, prediction, and experimentation, with the results of the experimentation causing possible modification of the hypothesis. Often, science is not carried out in as orderly a fashion as this indicates, but usually the sequence can be seen by historians at a later time. The scientific method is based on inductive, rather than deductive, reasoning; and, hence, there is no absolute certainty in science. The reason that the scientific method cannot work like a machine is that hypothesizing is a highly creative activity and cannot be programmed or scheduled. Furthermore, the scientific method is not easily applied to social science problems because most social science problems have too many variables and no laboratory in which to hold most variables constant.

Mathematics is useful to physical science because it is the collection of rules for dealing with quantities and the relationships between quantities. This book uses only simple algebra. Equations may be dealt with easily if you visualize them as physical balances, identify the unknown, and isolate it by performing the same operations to both sides of the equation. A variety of symbols are used in equations to represent physical quantities. Which symbol represents an unknown varies from one problem to another.

Numbers are usually expressed in standard scientific notation, which simplifies operating with very large or very small numbers.

Quantities that are directly proportional increase or decrease together.

A graph is a useful way to display some kinds of experimental results and to look for developing patterns.

Because quantities are so important to physical science, the way they are measured must be done carefully and systematically. Fundamental quantities are length, mass, and time. The metric system of units (MKS) defines the basic unit of length as the meter, the basic unit of mass as the kilogram, and the basic unit of time as the second. Each basic unit may be subdivided or multiplied by powers of ten by using a simple prefix before the unit: micro = 10^{-6}; milli = 10^{-3}; centi = 10^{-2}; kilo = 10^{3}; mega = 10^{6}.

Density provides an opportunity to practice math and unit analysis on a quantity which will be useful later. Density is the ratio of the mass of an object to its volume. For a given material, this ratio is independent of the amount of material you have, because the mass and volume are directly proportional to each other.

Two tricks sometimes show up in density problems: (1) no volume given, only dimensions (calculate the volume from a few simple formulas), and (2) dimensions mismatched (convert to metric units).

Finally, an experiment was used to demonstrate how the concept of density can be applied to a practical problem.

Questions

This may be used like a programmed text by referring to the indicated questions while reading the text. The answers are given in parentheses following the problems. (Note: One of the answers is incorrect. Perhaps you can find it.)

1.1

1. Trace the study of man and his environment by the social sciences, biological sciences, and physical sciences.
2. For what reasons did our earliest ancestors require some knowledge about the physical part of their environment?
3. What is natural philosophy, and how did man become able to study it?
4. What was the chief feature of the intellectual methods used by early natural philosophers?
5. What was Aristotle's theory about falling bodies?

1.2

6. In what respect was there a similarity between the changes in Western civilization as a whole and physical science as practiced in the West?
7. What was the major change in the method of studying natural philosophy?

8. What was Galileo's theory about falling bodies, and how did he substantiate it?
9. What sequence of activities is said to comprise the scientific method?
10. Give an example of how the scientific method might be applied to an everyday problem.
11. Explain how the "scientific method" is not always followed by scientists.
12. Explain the difference between inductive and deductive reasoning.
13. Give examples of inductive and deductive reasoning.
14. In what respect are scientific truths not "certain"?
15. Explain the unsuitability of the machine for illustrating the scientific method.
16. Explain why the scientific method has not been applied to all the problems of social science.
17. In what way does mathematics "fit" into the scientific method?
18. How do units systems "fit" into the scientific method?

1.3

19. (a) $v + 0.02 = 3.00$ ($v = 2.98$)
    (b) $c/12 = 7$ ($c = 84$)
    (c) $18\beta + 4 = 58$ ($\beta = 3$)
    (d) $11s = 44$ ($s = 4$)
    (e) $d - 8 = 62$ ($d = 70$)
    (f) $5f = 105$ ($f = 21$)
    (g) $g/3 + 8 = 21$ ($g = 39$)
    (h) $h + 16 = 12$ ($h = -4$)
    (i) $6e - 7 = 17$ ($e = 4$)
    (j) $m - 12 = 6$ ($m = 18$)
20. (a) $f\lambda = c$ Solve for λ (c/f)
    (b) $v = d/t$ Solve for t (d/v)
    (c) $a = \Delta v/\Delta t$ Solve for Δt ($\Delta v/a$)
    (d) $KE = \frac{1}{2} mv^2$ Solve for m ($2KE/v^2$)
    (e) $w = mg$ Solve for m (w/g)
    (f) $F = ma$ Solve for a (F/m)
    (g) $PE = mgh$ Solve for h (PE/mg)
21. Convert to standard scientific notation:
    (a) 652 (6.52×10^2)
    (b) 8.14 (8.14×10^0)
    (c) 0.000062 (6.2×10^{-5})
    (d) 43,000 (4.3×10^{-4})
    (e) 0.017 (1.7×10^{-2})
22. Convert from standard scientific notation to ordinary numbers:

(a) 6.12×10^{-6} (0.00000612)
(b) 8.1×10^{14} (810,000,000,000,000)
(c) 1.06×10^{0} (1.06)
(d) 9.9×10^{4} (99,000)
(e) 3.7×10^{-2} (0.037)

23. Convert to standard scientific notation and carry out the indicated operations:
(a) 612×2000 (1.224×10^{6})
(b) $4840 \div 20$ (2.42×10^{2})
(c) $0.00072 \div 8$ (9×10^{-5})
(d) 0.003×420 (1.26×10^{0})
(e) $0.0042 \div 0.000007$ (6×10^{2})
(f) 0.000081×0.003 (2.43×10^{-7})
(g) $0.00065 \div 5$ (1.3×10^{-4})
(h) $582 \div 0.00002$ (2.91×10^{7})

24. (a) $D = \frac{m}{V}$, $m = 12$; $D = 4$; $V = ?$ (3)
(b) $c = f\lambda$, $c = 3 \times 10^{8}$; $f = 6 \times 10^{5}$; $\lambda = ?$ (5×10^{2})
(c) $F = ma$, $a = 4$; $F = 68$; $m = ?$ (17)
(d) $v = \frac{d}{t}$, $v = 100$; $t = 0.01$; $d = ?$ (1)
(e) $a = \frac{\Delta v}{\Delta t}$, $\Delta t = 6$; $a = 100$; $\Delta v = ?$ (600)
(f) $w = mg$, $m = 4 \times 10^{-25}$; $g = 10$; $w = ?$ (4×10^{-24})
(g) $PE = mgh$, $m = 6$; $g = 10$; $h = 120$; $PE = ?$ (7.2×10^{3})
(h) $KE = \frac{1}{2}mv^2$, $m = 8$; $v = 10$; $KE = ?$ (400)
(i) $F_G = G\frac{m_1 m_2}{r^2}$, $F_G = ?$ (1.87×10^{-42})
$G = 7 \times 10^{-11}$; $m_1 = 4 \times 10^{-26}$; $m_2 = 6 \times 10^{-26}$; $r = 3 \times 10^{-10}$
(j) $F_E = K\frac{Q_1 Q_2}{r^2}$, $F_E = ?$ (2.56×10^{-9})
$K = 9 \times 10^{9}$; Q_1, $Q_2 = 1.6 \times 10^{-19}$; $r = 3 \times 10^{-10}$
(k) $E = hf$, $h = 6.6 \times 10^{-34}$; $f = 2 \times 10^{15}$; $E = ?$ (1.32×10^{-18})

25. Explain the notion of direct proportionality.
26. Give an example of two quantities that are directly proportional to each other.
27. Make a graph of the closing price of your favorite stock for the past week.

28. Make a graph of height vs. weight for a total of 10 people who are willing to tell you their height and weight.

1.4

29. What are the basic quantities which any units system must measure?
30. What are the basic units of the metric system?
31. What is the chief advantage of the metric system over the English system?
32. What are the principal decimal prefixes, and what do they stand for?
33. How many micrograms in a kilogram? (10^9)
34. How many megameters in a centimeter? (10^{-8})
35. How many milligrams in a megagram? (10^9)
36. How many kilometers in a megameter? (10^3)
37. How many kilograms in a microgram? (10^{-9})
38. How many millimeters in a centimeter? (10)

1.5

39. What is the difference between mass and volume?
40. Is the density of a small piece of lead the same as the density of a large piece of lead?
41. Solve the following density problems:

    (a) $m = 1000$ kg, $V = 200$ m^3; $D = ?$ $\left(5\,\frac{\text{kg}}{\text{m}^3}\right)$

    (b) $m = 0.01$ kg, $V = 2$ m^3; $D = ?$ $\left(5 \times 10^{-3}\,\frac{\text{kg}}{\text{m}^3}\right)$

    (c) $m = 10^5$ kg, $V = 4 \times 10^2$ m^3; $D = ?$ $\left(2.5 \times 10^{2}\,\frac{\text{kg}}{\text{m}^3}\right)$

    (d) $m = ?$ $V = 13$ m^3, $D = 10^3$ kg/m^3 (1.3×10^4 kg)
    (e) $m = ?$ $V = 3$ m^3, $D = 150$ kg/m^3 (450 kg)
    (f) $m = ?$ $V = 12$ m^3, $D = 7000$ kg/m^3 (8.4×10^4 kg)
    (g) $m = 12{,}000$ kg, $V = ?$ $D = 6000$ kg/m^3 (2 m^3)
    (h) $m = 6 \times 10^3$ kg, $V = ?$ $D = 3 \times 10^3$ kg/m^3 (2 m^3)
    (i) $m = 450{,}000$ kg, $V = ?$ $D = 150$ kg/m^3 (3000 m^3)
42. A cube of metal is 0.5 meter on a side. If its density is 3000 kg/m^3, what is its mass? (375 kg)
43. A cylinder of wood of density 500 kg/m^3 is one meter in radius and 10 meters long. What is its mass? (1.57×10^4 kg)
44. A metal cube has a density of 4 g/cc and a mass of 6 kg. What is its volume? (1500 cm^3)

45. 2000 kg of a heavy metal occupies a volume of 0.333 m^3. What is the metal's density in g/cm^3? (6 g/cc)

Quotes to Ponder

1. "The alarm rings as usual. You crawl out of bed, close the window, and check the outdoor thermometer: It's 10° Celsius. In the bathroom you step on the scales and discover that you've gained another kilogram—which means you've probably put on a few more centimeters around the middle, too. After bringing in two liters of milk from the back steps, you start to prepare breakfast—and begin another day with the metric system of weights and measures.

That day may not be far away. Last summer the National Bureau of Standards finished a three-year, $2.5 million 'metrication study.' As a result, there's now a bill before Congress that would provide for a 10-year conversion away from the familiar pounds, quarts, and inches to the more precise International System of Units. The metric system, which has its roots in the eighteenth century, is based on tens. There are, for instance, 1,000 millimeters or 100 centimeters in a meter. Water sensibly freezes at 0° and boils at 100°." (*McCalls,* March, 1972)

2. "The cost of an unplanned conversion to the metric system would be about one billion dollars for the chemical industry, over the duration of the program, MCA (Manufacturing Chemists Association) estimates.

Taking an international view, the U.S. is the only major country that has not yet committed itself to using the metric system. Currently, Great Britain is half-way through its 10 year plan for converting. Australia is also planning to go through a 10 year conversion plan and, to the north, Canada has declared a policy of eventual conversion. However, as NBS director Lewis M. Branscomb pointed out to the Deerfield conference, under some circumstances it is not only possible but desirable to permit the coexistence of more than one system of measurement language—as long as translation is possible. However, Dr. Branscomb asks: 'Will harmonizing our measurement language with other nations help mankind live in harmony with nature itself?'" (*Chemical & Engineering News,* August 24, 1970)

3. "Thus the experimental and mathematical methods were a growth, developing within the medieval system of scientific thought, which was to destroy from within and eventually burst out from Aristotelian cosmology and physics . . . There can be little doubt that it was the development of these experimental and mathematical methods of the 13th and 14th centuries that at least initiated the historical movement of the Scientific Revolution culminating in the 17th century." [Alistair C. Crombie, *The Continuity of Medieval and 17th Century Science*. From Basalla, G. (ed.) *The Rise of Modern Science* (paperback), D.C. Heath, Lexington, Mass., 1968]

4. "If Galileo Galilei were alive he probably would be tugging at his beard with pleasure about the elementary physics experiment Apollo 15 astronaut David R. Scott performed today on the moon.

Scott dropped a hammer and a feather from waist high to illustrate that both objects accelerated equally by the moon's gravity and that both would hit the surface at the same time despite their differences in mass or weight.

The experiment, similar to those Galileo did 300 years ago from the top of the Leaning Tower of Pisa in Italy, was performed as Scott and James B. Irwin were finishing up their final lunar excursion.

'In my left hand I have a feather. In the right hand a hammer,' Scott said, standing in front of the camera mounted on the lunar Rover.

'The reason I have these here today is because of Galileo's discourse on falling bodies in gravity fields. Where better to confirm his findings than on the moon.'

Then he dropped both objects and, sure enough, they struck the lunar surface simultaneously." (Associated Press News Release, July, 1971)

5. "Inch by ounce, a small group at Western Michigan University is planning a revolution for the United States. And it's being done with $163,214 of Uncle Sam's money.

But nobody's excited about it. That's because the revolution they are planning is an inevitable one—conversion of America to metric standards.

The Center for Metric Education at WMU is in its first month of a federally funded three-year program to prepare educational tools to help change this country's system of inches, ounces, feet, and so on to centimeters, grams, and meters.

The United States is now the only major nation not on

the metric system of measurement. Predictions are that this country will be 80 percent converted to metrics within the next decade.

That means learning a whole new language of meters, liters, and kilograms. WMU is working on educational materials to make that conversion as painless as possible.

Dr. John L. Feirer, head of the Department of Industrial Education at WMU, says it has been estimated that it will cost between $15 billion and $65 billion to convert America's industry to the metric system.

'The higher figure is usually quoted by opponents of the idea,' Dr. Feirer says. 'But judging from the experience in England, which converted recently, we believe the lower figure will be more accurate.'" (Associated Press News Release, November, 1972)

CHAPTER TWO

WHAT IS MATTER MADE OF?

The answer to this question, which is central to our entire study of matter, must certainly involve *atoms*. It would be easy enough to simply state the most modern understanding of an atom and demand that you memorize it, and then proceed to build up solids, liquids, and gases from various combinations of atoms. Instead, the most current model of an atom will be developed by looking at previous models of the atom as they evolved historically, making full use of observations, experiments, inductive and deductive reasoning.

A *model* is only a representation of some outside reality (e.g., a doll is a model of a human baby). Models can reflect the outside reality to varying degrees of accuracy, as in the case of rag dolls vs. molded plastic dolls vs. dolls with real hair, a voice, and functional legs and arms. Although the models of the atom that were developed historically were intellectual rather than physical such as a doll is, they have necessarily grown in complexity as they have attempted to portray reality more and more accurately.

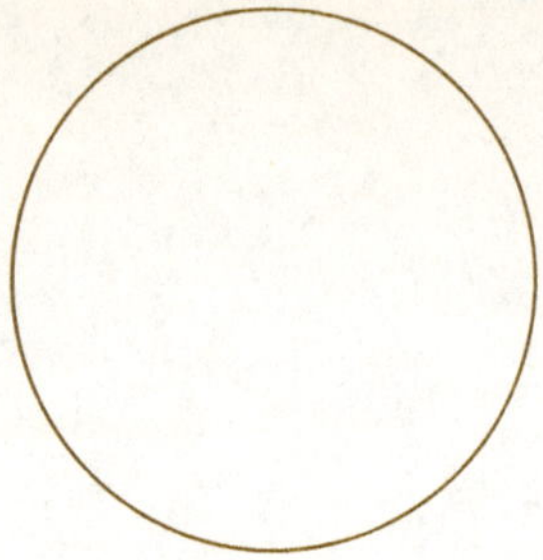

Figure 2.1 *Democritus' model of the atom*

2.1

Democritus' Model of the Atom

It has often been said that the ancient Greeks thought of everything first—atoms are no exception. Democritus (about 460 B.C.) was the first to record the speculation that there was some common feature underlying all matter, some basic structure from which all matter is made. He theorized (note the lack of scientific method) that the basic structure consisted of *tiny, indivisible, indestructible spheres* (Figure 2.1). They had to be tiny or else everyone could see them; they had to be indivisible because he postulated them to be the smallest particles in existence; they had to be indestructible because they had no parts; and they had to be spheres because the sphere is the most perfectly symmetrical three-dimensional shape, and the Greeks strongly believed in perfection. He named these structures *atoms,* which is Greek for uncutable or indivisible.

2.2

Thomson's Plum Pudding Model

After Galileo kicked off the first scientific revolution (there have been and will be other scientific revolutions), a whole new branch of scientific investigation was opened—*experimentation.* Of course, people had fooled around with various bits of apparatus before, but now that experiment played a central role, both in generating observations and

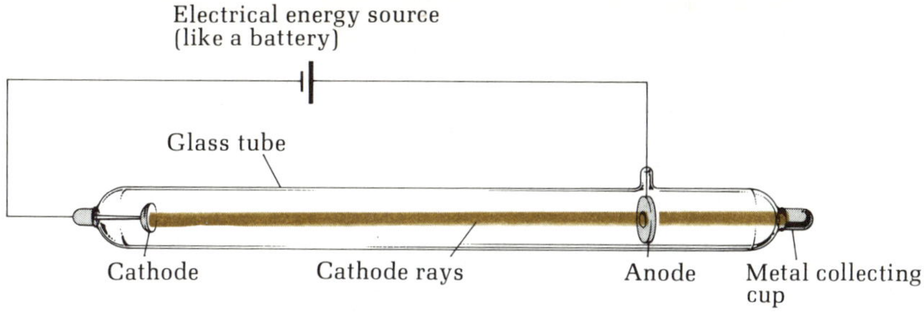

Figure 2.2 *Thomson's apparatus*

verifying predictions, scientists experimented with great glee, and stumbled onto many interesting things.

In the 1800's, glassblowing techniques were far enough advanced that many scientists conducted experiments using high quality blown glass apparatus. One example of this type of apparatus was the gas discharge tube used by Crookes, Faraday, Davy, Thomson, and others. Two electrical contacts and a metal collecting cup were sealed up in a glass tube along with a small amount of some gas, but no air. When the electrical contacts were hooked up to a battery, there seemed to be a "ray" inside the tube, running from the cathode toward the anode, as shown in Figure 2.2.

Thomson found that this ray caused the collecting cup to become negatively charged (*observation*).* He then formed the *hypothesis* that the ray consisted of negatively charged particles, and *predicted* that the cathode rays would be attracted toward positive charges. Thomson performed an *experiment* which verified this attraction of cathode rays toward a positively charged plate (see Figure 2.3).

Thomson named these negative charges *electrons* and measured the ratio of their charge to their mass:

$$\frac{\text{Electron charge}}{\text{Electron mass}} = 1.7 \times 10^{11}\ \frac{\text{coulombs}}{\text{kilogram}}.$$

It was later found that the charge of an electron is 1.6×10^{-19} coulombs, so its mass must be 9×10^{-31} kilograms. (Write this number out with the decimal point and all the zeros and you'll see why there is such a thing as standard scientific notation. Without it, we would lose zeros left and right.)

These electrons streamed out of the cathode regardless of

*Although we'll learn a lot more about the electric charge later, it is sufficient at this point to realize that charges are both positive and negative, like charges repel while unlike charges attract, and the unit of charge is the coulomb.

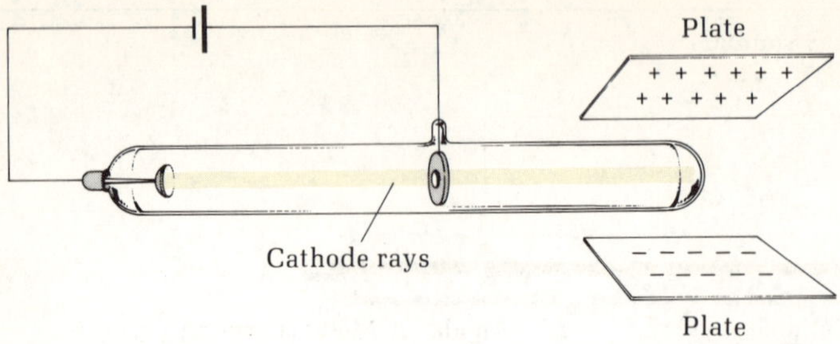

Figure 2.3 *Deflection of cathode rays toward positively charged plate*

the cathode material, so Thomson concluded that electrons must be a part of all atoms. (We've just begun to subdivide the atom.)

Since charges either attract or repel other charges and ordinary matter neither attracts nor repels other ordinary matter, we can conclude that most matter, and hence *atoms, have no net charge*.

Here, we can reason deductively:

(a) atoms contain no net charge;
(b) atoms contain electrons, which are negatively charged; therefore,
(c) atoms must contain as much positive charge as negative charge.

Ah, but where in the atom does the positive charge live? Thomson had no experimental evidence for the location of either the positive charge or the electrons, but he was willing to make a hypothesis.

Thomson envisioned the atom as a sphere of diameter about 10^{-10} meters with the positive charge distributed uniformly throughout, and the electrons scattered like raisins in a plum pudding, as shown in Figure 2.4. (You might prefer to call this model of the atom the matzoh ball soup or lumpy Cream-of-Wheat model. The point is simply to have mostly one substance with occasional chunks of another.) Notice that this second model of the atom was formulated from experimental results rather than theoretical deliberations.

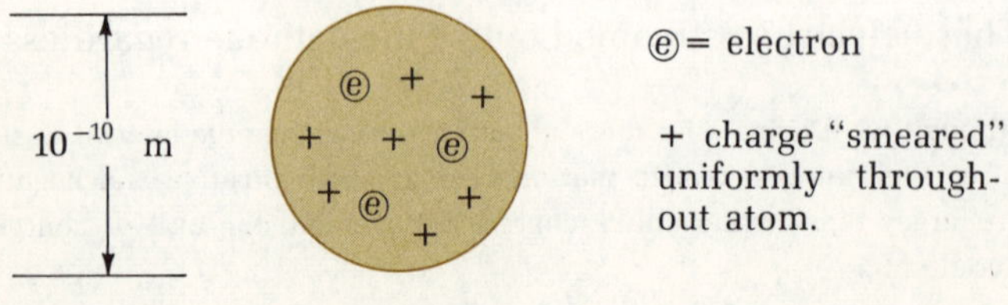

Figure 2.4 *Thomson's plum pudding model of the atom*

2.3

Rutherford's Solar System Model

No unverified hypothesis exists in science for very long, because there are plenty of eager experimentalists. In the case of Thomson's plum pudding model of the atom, however, there were certain difficulties. Let's look at the size of the atom for a moment and we'll see why experiments were so tough to design.

If an atom is really 10^{-10} meters in diameter, then how many atoms are there in the thickness of this page? Well, books go about 100 pages to the centimeter, so each page must be 1/100 centimeter or 10^{-2} centimeter, or 10^{-4} meters in thickness. How many 10^{-10} diameter atoms will fit in this thickness?

$$\frac{10^{-4} \text{ meters}}{10^{-10} \text{ meters/atom}} = 10^{-4-(-10)} \text{ atoms} = 10^{6} \text{ atoms.}$$

One page is a million atoms thick! That doesn't sound like too much. Corporations and politicians can easily spend a million dollars these days, but most of us have difficulty visualizing numbers this large.

10^6 small ants would form a line 7 blocks long;
10^6 books this size stacked one atop another would be three times the height of the tallest mountain;
10^6 people standing shoulder to shoulder would stretch across almost any state;
10^6 automobiles bumper to bumper would stretch all the way across the country (what a traffic jam!).

By now, it should be clear that because of the atom's size, scientists had no way of getting into one single atom to see what it was made of. Don't give up too easily, though, as the next experiment shows; even if you can't see inside something, you can get a pretty good idea of what's inside by using less direct means.

EXPERIMENT 2.1
BLACK BOX

Just as the scientists in the 1800's couldn't see what was inside individual atoms, you won't be able to see what is inside the black boxes (provided by your instructor).

Your lab instructor may give you some additional facts about the contents of the box, but it will be your job to form teams, decide which rods to pull out and in what order, and generally plan a systematic approach to finding out the number and placement of objects within the box. (Hint: Try the old scientific method—what have you got to lose?) Now you know a little of what the scientists had to go through to find out what was inside an atom without actually seeing inside it.

Of course, there are a number of ways to explore something you can't see. Suppose you wanted to explore a distant haystack but couldn't sort through it straw by straw. If you had a cannon and a large enough supply of cannonballs, you could shoot the cannonballs through the haystack and see what happened to the cannonballs, as shown in Figure 2.5. The problem of the 19th century scientists was to find atomic-sized cannonballs.

To the rescue came another strange phenomenon—*radioactivity.* You may have heard the old tale about the scientist, Becquerel, who put an unusual rock on top of a key on top of an unexposed photographic plate in a drawer. The photographic plate was exposed later and found to have an image of the key on it. The rock must have been giving off some kind of ray, and Becquerel, the Curies, and other now-famous persons attempted to find out what the strange rays were.

Adopting the same technique as Thomson did with cathode rays, the scientists ran the radioactive ray between two charged plates to see which way it would go when it lit up the phosphor screen. Much to their chagrin, it split into

Figure 2.5 *Haystack exploration by cannonballs*

Figure 2.6 *Apparatus for exploring radioactivity*

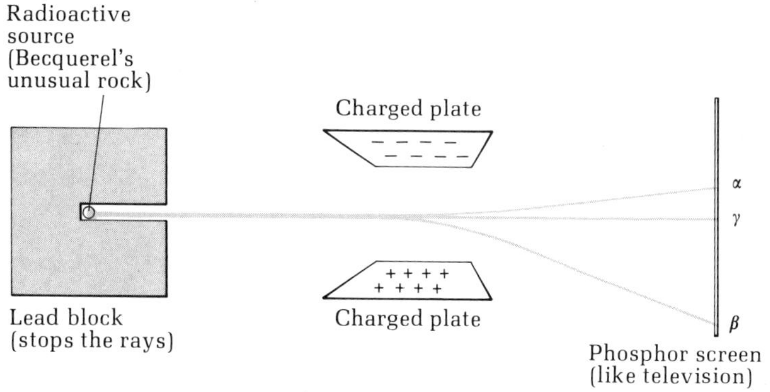

three parts, as shown in Figure 2.6. Before, they had only one ray to deal with; now there were three.

The three rays were named α (alpha), β (beta), and γ (gamma), the first three letters in the Greek alphabet. The α was attracted by the negatively charged plate, and its path did not deviate much from a straight line; so it must be positively charged and fairly massive. The β was attracted by the positively charged plate and deflected greatly; so it must be negatively charged and not too massive. The γ just continued undeflected; so it must be uncharged. Although scientists were not at all certain about these things at the time, the γ's are very similar to visible light, the β's are electrons, and the α's are the nuclei of helium atoms. What scientists *did* realize was that the relatively massive, positively charged α particles could be just the atomic cannonballs needed to find the location of the positive charge within the atom.

Thomson's successor, Rutherford, and his graduate assistants, Geiger and Marsden, used α particles in an experimental attempt to verify Thomson's model of the atom. Their apparatus, shown schematically in Figure 2.7, bombarded an extremely thin sheet of gold foil with α particles, while they watched the phosphor screen to see how many α particles got through to light up the screen.

Their prediction, based on Thomson's model of the atom, was that most small α particles would go right through the larger positive charges in the atom with only a slight deflection due to repulsion between the two + charges. Sure enough, their initial experiments seemed to bear this out. When they put the gold foil in the beam, the beam seemed to grow only slightly fuzzy at the edges.

Marsden was given the job of checking to see if any α par-

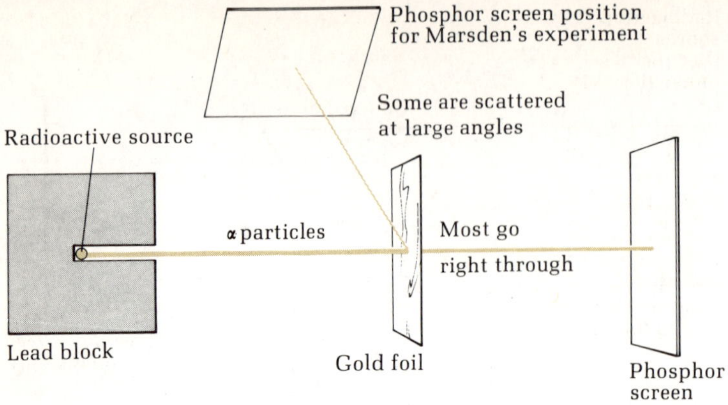

Figure 2.7 *Rutherford, Geiger, and Marsden's apparatus for bombarding gold atoms with α-particles*

ticles were deflected at large angles. Very little deflection was expected, but it would keep Marsden busy for quite a while because the screen was only expected to flash a few times a day. It turned out that the phosphor screen flashed considerably more than that, leading to the conclusion that something was wrong with Thomson's model. The only way to make the model fit the observations was to put the positive charge into a small volume within the atom. A single positive charge was believed to be a single particle, and was called a *proton*. The place where the protons were concentrated into the center of the atom was called the *nucleus*. The electrons were thought to be orbiting outside the nucleus, like planets; hence the solar system name for the model. This third model of the atom is sketched in Figure 2.8.

If we take the simplest atom—hydrogen—consisting of a single proton and a single electron, and take away its electron, we find that the mass of the proton is almost 2000 times the mass of the electron; however, the proton's charge has the same magnitude but is positive, while the electron's charge is negative. The mass ratio between the proton and the electron may be illustrated by the following equation:

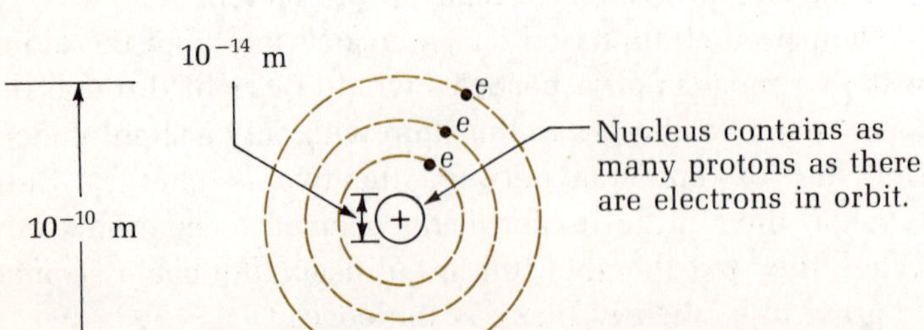

Figure 2.8 *Rutherford's solar system model of the atom*

$$\frac{\text{Proton mass}}{\text{Electron mass}} = \frac{\text{Bowling ball}}{\text{Quarter}} = \frac{\text{Man}}{\text{Mouse}} = \frac{\text{Building}}{\text{Man}}$$

2.4

Rutherford's Model Refined—The Neutron

As man's understanding of the properties of elements grew by means of experiments, the model of the atom had to be revised to reflect reality. The next change involved a new particle in the nucleus besides the proton—the *neutron.* Although it may look roundabout, the way to the neutron starts with some chemistry and the periodic table.

An English schoolteacher and chemist, John Dalton, organized a list of the atoms in the early 1800s. He and the other chemists of that era arranged chemical reactions between elements (groups of the same kind of atoms) and were able to obtain the masses of each of the known atoms relative to each other. For example, by causing a chemical reaction between hydrogen and oxygen to form water, they found that the oxygen atom is 16 times as massive as the hydrogen atom. By causing a chemical reaction between magnesium and oxygen to form magnesium oxide, they found that an atom of magnesium is 1.5 times as massive as an oxygen atom; hence $(1.5)(16) = 24$ times as massive as a hydrogen atom. Chemists, being what they are, caused every conceiv-

Table 2.1 *Relative Masses of Atoms*

ATOM	MASS	ATOM	MASS
Hydrogen	1.0	Sodium	23.0
Helium	4.0	Magnesium	24.3
Lithium	6.9	Aluminum	27.0
Beryllium	9.0	Silicon	28.1
Boron	10.8	Phosphorous	31.0
Carbon	12.0	Sulfur	32.1
Nitrogen	14.0	Chlorine	35.5
Oxygen	16.0	Potassium	39.1
Fluorine	19.0	Argon	40.0
Neon	20.2	Calcium	40.1

able chemical reaction, and compiled a list of elements arranged in order of increasing mass. This list was based on the hydrogen atom, the simplest of all atoms, as the standard mass. Table 2.1 gives some elements from that list.

A Russian chemist, Dmitri Mendeleev, noticed that there were similarities in the chemical properties of many of the elements; so he proposed a new grouping of the elements, putting elements with similar chemical properties in a vertical column.

Mendeleev suggested listing the elements horizontally, in their order of increasing atomic mass; when an element with chemical properties similar to the ones listed was found, he started a new row (see Table 2.2). For example, lithium, sodium, and potassium are all highly reactive metals that react in the same proportions with fluorine, chlorine, and so forth; so lithium, sodium, and potassium each begin a

Table 2.2 *Periodic Table of the Elements*

1 H 1.0																	2 He 4.0
3 Li 6.9	4 Be 9.0											5 B 10.8	6 C 12.0	7 N 14.0	8 O 16.0	9 F 19.0	10 Ne 20.2
11 Na 23.0	12 Mg 24.3											13 Al 27.0	14 Si 28.1	15 P 31.0	16 S 32.1	17 Cl 35.5	18 Ar 39.9
19 K 39.1	20 Ca 40.1	21 Sc 45.0	22 Ti 47.9	23 V 50.9	24 Cr 52.0	25 Mn 54.9	26 Fe 55.8	27 Co 58.9	28 Ni 58.7	29 Cu 63.5	30 Zn 65.4	31 Ga 69.7	32 Ge 72.6	33 As 74.9	34 Se 79.0	35 Br 79.9	36 Kr 83.8
37 Rb 85.5	38 Sr 87.6	39 Y 88.9	40 Zr 91.2	41 Nb 92.9	42 Mo 95.9	43 Tc (97)	44 Ru 101.1	45 Rh 102.9	46 Pd 106.4	47 Ag 107.9	48 Cd 112.4	49 In 114.8	50 Sn 118.7	51 Sb 121.8	52 Te 127.6	53 I 126.9	54 Xe 131.3
55 Cs 132.9	56 Ba 137.3	see below 57-71	72 Hf 178.5	73 Ta 180.9	74 W 183.9	75 Re 186.2	76 Os 190.2	77 Ir 192.2	78 Pt 195.1	79 Au 197.0	80 Hg 200.6	81 Tl 204.4	82 Pb 207.2	83 Bi 209.0	84 Po 210	85 At (210)	86 Rn (222)
87 Fr (223)	88 Ra (226)	see below 89-103															

57 La 138.9	58 Ce 140.1	59 Pr 140.9	60 Nd 144.2	61 Pm (145)	62 Sm 150.4	63 Eu 152.0	64 Gd 157.3	65 Tb 158.9	66 Dy 162.5	67 Ho 164.9	68 Er 167.3	69 Tm 168.9	70 Yb 173.0	71 Lu 175.0
89 Ac (227)	90 Th 232.0	91 Pa (231)	92 U 238.0	93 Np (237)	94 Pu (244)	95 Am (243)	96 Cm (247)	97 Bk (247)	98 Cf (251)	99 Es (254)	100 Fm (257)	101 Md (256)	102 No (254)	103 Lw (257)

new row. Another example is the sequence beginning with helium, neon, argon. These are gases which are almost completely inert chemically, and thus are called "noble gases" because they are "above" participating in chemical reactions. Another example is the sequence of fluorine, chlorine, bromine, and so forth. These are called the halogens and also have very similar chemical properties.

Anyway, this suggestion of Mendeleev's proved quite successful because there were several blanks in the list of elements, and he boldly predicted the mass and chemical properties of the "missing" elements. They were later discovered and fit into his scheme very nicely. (See if you can trace the "scientific method" in this intellectual development.)

The modern version of Mendeleev's scheme is called the *Periodic Table of the Elements*, shown in Table 2.2.

To the scientist, this periodic table is extremely meaningful. The elements are like his friends. They all have their own personalities, and the periodic table is like some kind of organization chart or a seating chart for some atomic banquet. Let's analyze the chart a little closer.

The Symbols. Each symbol is just a shorthand form of the element name. It generally consists of the first one or two letters (with some exceptions) in the Latin or English word for the element. The first letter is capitalized and the second is small. For example:

H   Hydrogen (first letter)
He   Helium (first two letters)
Li   Lithium (first two letters)
⋮
Na   Sodium (Latin name *natrium*)
⋮
K   Potassium (Latin name *kalium*)
⋮
Ag   Silver (Latin name *argentum*, but Ar is used for Argon; therefore Ag is used for silver)

Clearly, the point of these symbols is to avoid having to write out the long names of elements too often. (Chemists don't like to write long words repeatedly any more than the rest of us do.) To avoid confusion, no two elements have the same symbol.

Table 2.3 lists the element names in alphabetical order with their corresponding symbols.

Table 2.3 *Elements and Their Symbols*

ELEMENT	SYMBOL	ELEMENT	SYMBOL
Actinium	Ac	Mercury	Hg
Aluminum	Al	Molybdenum	Mo
Americium	Am	Neodymium	Nd
Antimony	Sb	Neon	Ne
Argon	Ar	Neptunium	Np
Arsenic	As	Nickel	Ni
Astatine	At	Niobium	Nb
Barium	Ba	Nitrogen	N
Berkelium	Bk	Nobelium	No
Beryllium	Be	Osmium	Os
Bismuth	Bi	Oxygen	O
Boron	B	Palladium	Pd
Bromine	Br	Phosphorus	P
Cadmium	Cd	Platinum	Pt
Calcium	Ca	Plutonium	Pu
Californium	Cf	Polonium	Po
Carbon	C	Potassium	K
Cerium	Ce	Praseodymium	Pr
Cesium	Cs	Promethium	Pm
Chlorine	Cl	Protactinium	Pa
Chromium	Cr	Radium	Ra
Cobalt	Co	Radon	Rn
Copper	Cu	Rhenium	Re
Curium	Cm	Rhodium	Rh
Dysprosium	Dy	Rubidium	Rb
Einsteinium	Es	Ruthenium	Ru
Erbium	Er	Samarium	Sm
Europium	Eu	Scandium	Sc
Fermium	Fm	Selenium	Se
Fluorine	F	Silicon	Si
Francium	Fr	Silver	Ag
Gadolinium	Gd	Sodium	Na
Gallium	Ga	Strontium	Sr
Germanium	Ge	Sulfur	S
Gold	Au	Tantalum	Ta
Hafnium	Hf	Technetium	Tc
Helium	He	Tellurium	Te
Holmium	Ho	Terbium	Tb
Hydrogen	H	Thallium	Tl
Indium	In	Thorium	Th
Iodine	I	Thulium	Tm
Iridium	Ir	Tin	Sn
Iron	Fe	Titanium	Ti
Krypton	Kr	Tungsten	W
Lanthanum	La	Uranium	U
Lawrencium	Lw	Vanadium	V
Lead	Pb	Xenon	Xe
Lithium	Li	Ytterbium	Yb
Lutetium	Lu	Yttrium	Y
Magnesium	Mg	Zinc	Zn
Manganese	Mn	Zirconium	Zr
Mendelevium	Md		

The Numbers. The *smaller* of the two numbers in each box in the periodic table is called the *atomic number.* Envision it as being arrived at experimentally by taking electrons away from an atom, and simply using the number of electrons as the atomic number. Because of the electrically neutral nature of atoms, the number of electrons must be balanced by an equal number of protons:

Smaller number = atomic number = number of electrons
= number of protons

The *larger* of the two numbers in the periodic table is called the *atomic mass,* and may be arrived at experimentally by reacting elements together while keeping track of the masses involved. (This is where the neutron comes in, so wait two pages for a full explanation of the atomic mass.)

Elements should be listed with the symbol, the atomic mass, and the atomic number in the following form: $^{4}_{2}He$, helium, atomic mass 4 and atomic number 2; $^{27}_{13}Al$, aluminum, atomic mass 27 and atomic number 13.

Two terrible problems arise when one starts getting into atoms using the periodic table and Rutherford's atomic model:

1. *Not all atoms seem to have a whole number of protons.* Look at lithium, $^{6.9}_{3}Li$. The atomic mass is 6.9, so one would expect it to have 6.9 protons. Chlorine, $^{35.5}_{17}Cl$, has an atomic mass of 35.5; so it should have half a proton running around somewhere inside its nucleus.
2. *Protons belonging to different atoms seem to have different masses.* Ignoring electrons, because it takes almost 2000 of them to make one hydrogen atom, let's look at some masses.

The hydrogen atom, $^{1}_{1}H$, has 1 electron and 1 proton and its total mass is 1; so the proton mass is 1. The helium atom, $^{4}_{2}He$, has 2 electrons and 2 protons and a total mass of 4; so each proton has a mass of 2. Lithium, $^{6.9}_{3}Li$, has 3 electrons and 3 protons and a mass of 6.9; so each proton has a mass of 2.3. The protons seem to have different masses, depending on which atom they are in. This is bad, because we originally expected protons to be universal, not highly variable.

There could be a number of ways out of this quandary, but the one way which proved successful was the invention of a new particle—the *neutron.* To balance the mass without upsetting the charge balance, the neutron must have mass equal to the proton, but it must have no charge.

Let's see how this new particle, the neutron, solves the difficulties we saw previously. First of all, how many neutrons should there be? Just enough to add to the protons to obtain the mass which has been determined experimentally. For example, in helium, $^{4}_{2}He$, to get to a mass of 4 and have the necessary 2 protons, there must be 2 neutrons.

As another example, beryllium, $^{9}_{4}Be$, must have 4 protons because of its 4 electrons; so to get to a mass of 9, it must have 5 neutrons.

How do neutrons account for the fractional numbers in the periodic table? Since neutrons have no charge, there are no electrical requirements governing the number of neutrons which may be in the nucleus. Thus, there may be several different possibilities for the number of neutrons in the nucleus. As an example, there are two "different" chlorine atoms, $^{35}_{17}Cl$ and $^{37}_{17}Cl$. Both chlorine atoms have 17 electrons and 17 protons, but one has 18 neutrons and one has 20 neutrons in its nucleus. Although the nuclei are different, the electrons, which determine the chemical properties, are arranged precisely the same way; so the two kinds of chlorine atoms are ordinarily chemically indistinguishable. Thus, ordinary chlorine is a mixture of the two kinds (about 75% of the 18 neutron, 17 proton type, and 25% of the 20 neutron, 17 proton type), and so the atomic mass of 35.5 is actually an average of the two kinds. These different varieties of the same atom are called *isotopes,* which is Greek for "same spot," because they occupy the same spot in the periodic table. So, the theoretical invention of the neutron has bailed the atomic model out of an uncomfortable spot. The neutron's existence was not demonstrated until 1932; so you can imagine how elusive it was.

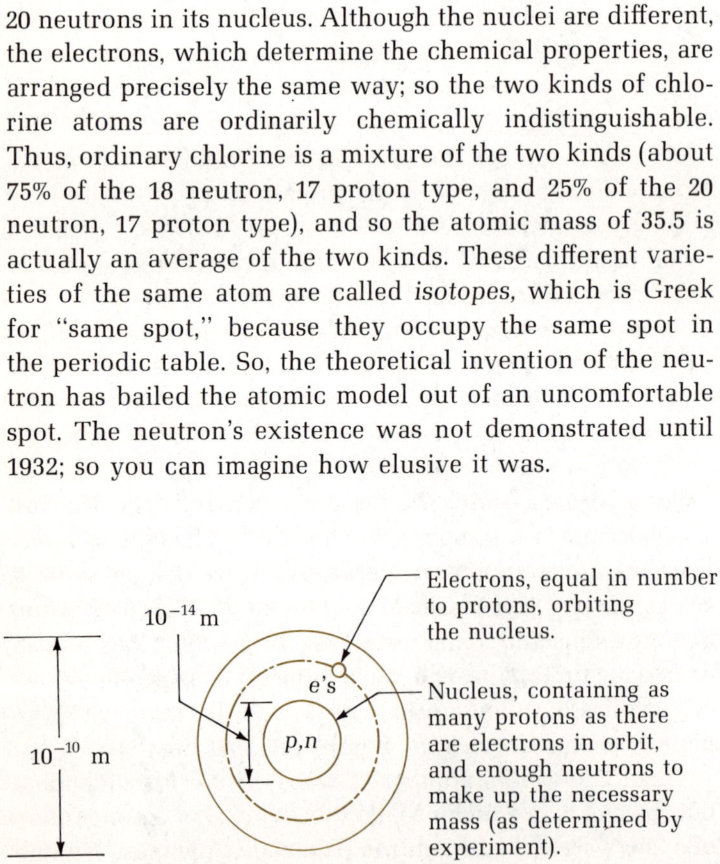

Figure 2.9 *Rutherford's model refined*

Now, let's go back and make sure we have the atomic mass properly figured out.

Larger number = atomic mass
= number of protons + number of neutrons

For example, phosphorus, ${}^{31}_{15}P$, has an atomic number of 15 and an atomic mass of 31; so it has 15 electrons, 15 protons, and 16 neutrons.

Our new model of the atom is illustrated in Figure 2.9.

2.5

Arrangement of Electrons— The Energy Level Model

The neutrons and protons in the nucleus correspond to our most modern model of the atom, but the idea of electrons orbiting the nucleus, like planets orbit the sun, needs some modification. First of all, let's see if there is any pattern to the arrangement of electrons.

We'll study the electron arrangement by either adding or taking away electrons from atoms. Atoms which have gained or lost electrons are now charged and are, therefore, no longer atoms; they are called *ions*. This process of adding or taking away electrons is called *ionization*.

First, let's look at the results of experiments in which a *single electron* was taken from each atom, in order of increasing atomic number. Each atom "holds onto" its electrons with varying amounts of strength; the experiments measured the energy necessary to take away a single electron from each atom. We'll explore the concept of energy in great detail in Chapter 5, but, for now, just think of energy as a measure of the "tightness" with which an atom holds onto its electrons. Figure 2.10 shows the results of a series of experiments taking a single electron from each atom.

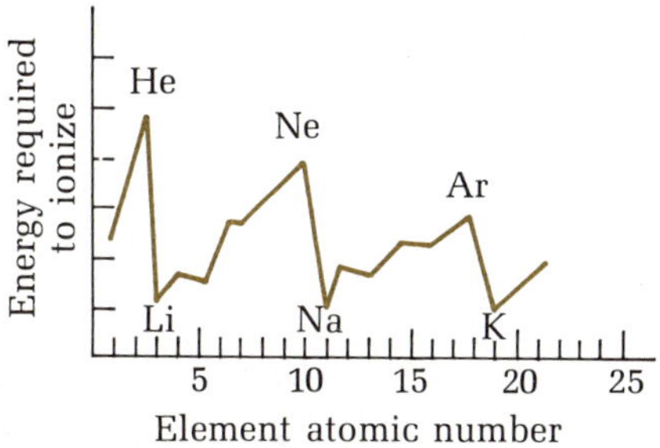

Figure 2.10 *Energy required to take a single electron from each atom*

Notice that lithium (Li), sodium (Na), and potassium (K) give up an electron with relative ease. These elements are alkali metals and occupy a vertical column in the periodic table. Notice that helium (He), neon (Ne), and argon (Ar) each hold their electron very tightly. These three elements are noble gases and occupy a vertical column in the periodic table. If you are beginning to suspect that electron arrangement and chemical properties are directly related, you are on the right track.

Next, let's look at the results of an experiment in which all electrons are taken from an atom, one at a time. Again, the energy required to remove each electron may be thought of as a measure of the "tightness" with which the electron is held by the atom. The results of this experiment are shown in Figure 2.11 for the atoms of boron and carbon.

Of boron's 5 electrons, the first 3 come off relatively easily, while the last 2 come off with great difficulty. Of carbon's 6 electrons, the first 4 come off relatively easily, while the last 2 again come off with great difficulty. There is a distinct pattern developing here. When this experiment was carried out for other atoms, the last 2 electrons were always the most difficult to get off. This led scientists to postulate that electrons were arranged in distinct *energy levels* within atoms, as shown in Table 2.4. Although you might think this

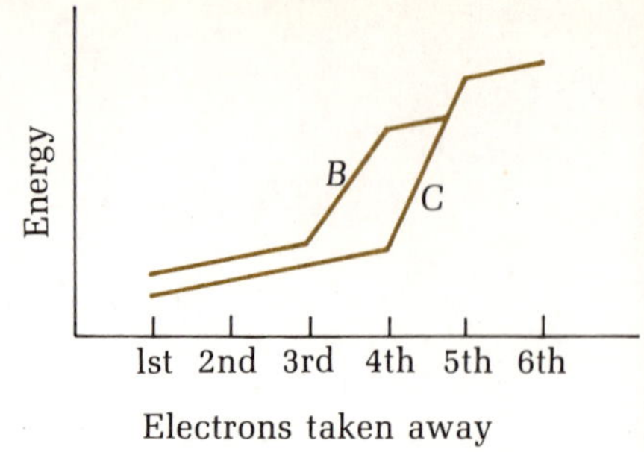

Figure 2.11 *Energy necessary to take all electrons from boron and carbon*

Table 2.4 *Electron Arrangements within Atoms*

Atom		Number of Electrons: First energy level	Second energy level	Third energy level	Fourth energy level
Hydrogen	H	1			
Helium	He	2			
Lithium	Li	2	1		
Beryllium	Be	2	2		
Boron	B	2	3		
Carbon	C	2	4		
Nitrogen	N	2	5		
Oxygen	O	2	6		
Fluorine	F	2	7		
Neon	Ne	2	8		
Sodium	Na	2	8	1	
Magnesium	Mg	2	8	2	
Aluminum	Al	2	8	3	
Silicon	Si	2	8	4	
Phosphorus	P	2	8	5	
Sulfur	S	2	8	6	
Chlorine	Cl	2	8	7	
Argon	Ar	2	8	8	
Potassium	K	2	8	8	1
Calcium	Ca	2	8	8	2

pattern looks very nice and regular to this point, the electron energy levels are actually more complicated than this indicates, particularly at larger atomic masses.

Notice that there are never *more* than 2 electrons in the first energy level and never *more* than 8 electrons in the second energy level. This led to the notion of *"filled" energy levels,* which explains ions and chemical reactions beautifully.

Recall that an ion is simply an atom that has gained or lost one or more electrons and hence become charged. Chemists have determined experimentally the number of electrons gained or lost when atoms become ions. A few examples of ions and the number of electrons gained or lost to form these ions are given in Table 2.5.

Now, if we look at the number of electrons gained or lost compared to the number of electrons necessary to "fill" an energy level, we'll see a direct relationship. For example, lithium loses the single electron in its second energy level to become an ion and thus it has only a filled energy level remaining. Fluorine gains one electron, which fills its second energy level with 8. Thus, it appears that *atoms form ions by gaining or losing enough electrons to fill an energy level.*

The filled energy level concept also explains many chemical reactions in terms of electrons being transferred from one atom to another in just the right number to fill each atom's energy levels. For example, it has been experimentally determined that calcium and chlorine combine to form a salt, calcium chloride, in the ratio of one atom of sodium to two atoms of chlorine. The chemical formula is

Table 2.5 *Some Ions and the Number of Electrons Required to Form Them*

Ion		Number of Electrons Which the Atom Has: Gained	Lost
Lithium	Li^{+1}		1
Beryllium	Be^{+2}		2
Fluorine	F^{-1}	1	
Sodium	Na^{+1}		1
Magnesium	Mg^{+2}		2
Chlorine	Cl^{-1}	1	
Potassium	K^{+1}		1

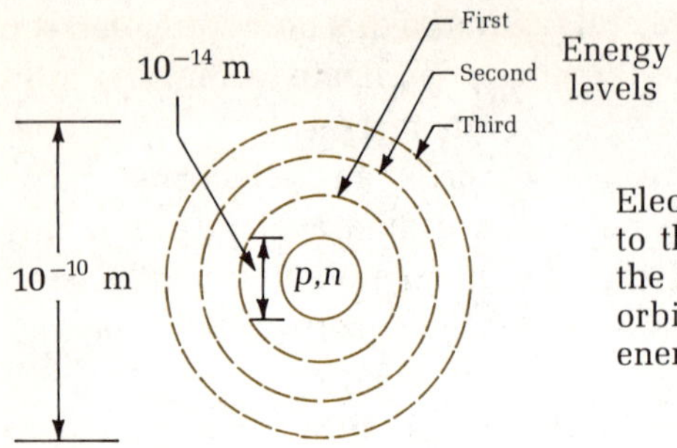

Figure 2.12 *Energy level model of the atom*

written $CaCl_2$ where the subscript 2 on the Cl refers to 2 atoms of Cl, and the missing subscript must be understood as a 1 for 1 atom of Ca. Calcium forms its common ion Ca^{+2} by losing 2 electrons, while a chlorine atom must gain a single electron to become an ion. Thus, if the calcium atom loses two electrons, and two chlorine atoms each gain a single electron, all three become ions, all three have filled energy levels, and no electrons are running around loose.

Many other chemical reactions may be explained in a similar way, with electrons being transferred to fill energy levels. Scientists extended the energy level model of the atom to say that the electrons had definite orbits, with the first energy level corresponding to an orbit close to the nucleus, the second energy level corresponding to an orbit a little farther out. Thus, the electrons which are the farthest away from the nucleus would be the easiest to take away from the atom, and the closer the orbit, the harder it would be to take the electrons. This energy level model is shown in Figure 2.12.

The energy level model of the atom has a basic flaw in that it cannot explain a fundamental property of atoms—radiation.

2.6

The Bohr Model of the Hydrogen Atom—Radiation Explained

It is not difficult to demonstrate the fact that atoms radiate light. Take a glass tube (slightly bent up), remove the air, put in a little neon ${}^{20}_{10}Ne$ gas, put electrodes at both ends,

Figure 2.13 *Radiation from neon atoms*

turn on the electric current, and you have . . . Figure 2.13.

Neon signs constantly give off their garish red-orange radiation, encouraging us to buy or consume various things. Yet how can we use the energy level model to explain why the neon atom gives off light, and, specifically, why does it *always* give off the same color of light? Other signs which contain other gases give off different colors, but each gas always gives off the same color.

To modify our model of the atom in order to explain radiation, we'll have to backtrack just a bit and look at a theoretical model developed to explain radiation from heated bodies. Around 1900, Rayleigh and Jeans set out to analyze the energy radiated by heated bodies. They presumed that a heated body consisted of a collection of small "radiators," each of which could have any amount of energy. When Rayleigh and Jeans summed up the energy of all the radiators, they were embarrassed to find that their theory predicted that the total energy being radiated is *infinite*. Of course, the total energy is finite, and their result has been termed "the ultraviolet catastrophe." (Who says scientists don't have a sense of humor?) The failure of their theory to make the correct prediction was published, and other scientists inspected the theory to see how it could be remedied.

Max Planck suggested that the difficulty could be cured if the assumption that the radiators could have *any* energy was changed to say that the radiators could have only particular amounts of energy. These particular amounts of energy all had to be whole number multiples of a minimum size packet of energy called a *quantum* of energy. This was the beginning of the notion of discrete chunks of energy rather than the old notion that energy could come in any size. The difference between coming in any size or only in multiples of a certain minimum size is illustrated by continuous vs. link sausage in Figure 2.14. You can buy the continuous form of sausage in any length you request, but link sausage is normally purchased in some whole number of links.

Not only did the *quantum* of energy solve the problem of radiation from heated bodies, but it was also applied by Bohr to the hydrogen atom in 1913. Bohr's model of the hydrogen atom hypothesized that hydrogen's single electron could only go around the nucleus in certain "allowed" orbits and, therefore, the electron had only particular "allowed" energy levels, and the theory predicted what these energy levels were. Furthermore, Bohr's model stated that an atom would *absorb* energy by having its electron go from some lower energy level corresponding to an inner orbit to an outer orbit corresponding to a higher energy level. The atom would then *give off* the energy it had absorbed by having its electrons jump from the higher energy level to some lower energy level. The energy given off would be in the form of radiation, and the frequency of the radiation would be directly proportional to the energy lost according to the relation:

$$\Delta E = hf.$$

Where ΔE is the difference in energy levels from where the electron started to where it ends, the energy unit is the *joule* (you'll see just where energy and its unit comes from in Chapter 5); h is a constant, known as Planck's constant, and

Figure 2.14 *Unlike energy, sausage comes in two forms, discrete and continuous*

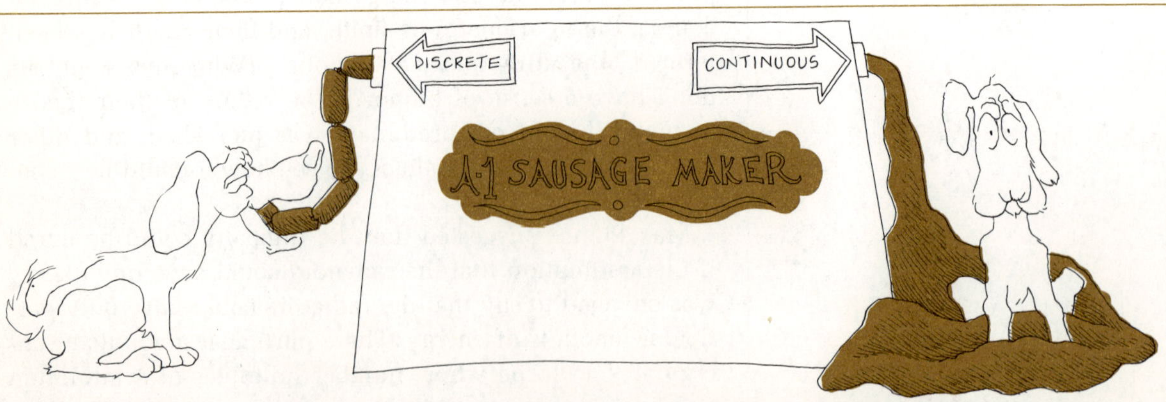

Figure 2.15 *T. H. Ump demonstrates different energy levels*

its value is 6.6×10^{-34} joule-sec/cycle; f is the frequency (cycles/sec) of the radiation that is given off.

To get another view of this energy level jumping business in atoms, let's look at an analogous situation portrayed in Figure 2.15 by another member of the Ump family, T. H. Ump.

T. H. Ump can absorb enough energy to get up to any level above him, but he must land *on* a level, and cannot land somewhere between levels. When T. H. Ump comes back down, he can jump from whatever level he is on to any level below, but he must jump from one level to another lower level and lose some energy in the process.

Hydrogen's electron behaves the same way as T. H. Ump, absorbing energy as it jumps from a lower energy level to a higher energy level, then the electron jumps back down to the lowest level in any number of jumps as long as the jumps are from one level to another. When the electron makes a jump to a lower energy level, the difference in energy is given off as radiation, according to the Planck relation; that is,

$$\Delta E = hf.$$

Let's try an experiment to see if the Bohr model of the hydrogen atom really works.

EXPERIMENT 2.2
HYDROGEN SPECTRUM ANALYSIS

To check Bohr's theory, we'll assemble a group of hydrogen atoms in a vacuum tube, supply them with enough electrical energy so that their electrons should jump up to higher energy states, then watch for the radiation the electrons should give off as they come back down. Since there are lots of atoms, and we're supplying plenty of energy, electrons should be jumping up to every upper energy state and then back down to every lower energy state, giving off many different frequencies of radiation in the process.

A. Observation of the Radiation

When the hydrogen-filled tube lights up, the radiation you see is a mixture of many different frequencies. First of all, we need to separate the light into its components. We can split the light into components using a device called a spectrograph, which usually shows the different colors in the radiation superimposed on a scale which gives the wavelength directly. You should see 4 lines: red, blue-green, blue, and violet. Record the wavelengths of the lines you see.

Example #1, violet:

$$4.0 \times 10^{-5} \text{ cm/cycle} = 4.0 \times 10^{-7} \text{ m/cycle}.$$

B. Conversion from Wavelength to Frequency

In the next chapter you will learn all the details about the relationship between frequency, wavelength, and speed of light. At this point, simply use the relationship: $f\lambda = c = 3 \times 10^8$ m/sec. The wavelength, λ, was determined in step A above, so find the frequency, f, for each of the lines you observed.

$$f = \frac{c}{\lambda} = \frac{3 \times 10^8 \text{ m/sec}}{\lambda}.$$

Example #2, violet:

$$\lambda = 4.0 \times 10^{-7} \text{ m/cycle}.$$

$$f = \frac{3 \times 10^8 \text{ m/sec}}{4.0 \times 10^{-7} \text{ m/cycle}} = 7.5 \times 10^{15} \text{ cycles/sec}.$$

C. Calculation of Energy Level Difference

Each of these lines you see is supposed to have arisen because an electron made a jump from a higher energy level to a lower energy level. The relationship between the frequency observed and the energy level difference is $\Delta E = hf$ where $h = 6.6 \times 10^{-34}$ joule-sec/cycle.

Calculate ΔE for each of the lines you observed.

Figure 2.16 *"Allowed" energy levels for hydrogen*

	ENERGY (Joules)
6th Bohr orbit	21.2×10^{-19}
5th Bohr orbit	21.0×10^{-19}
4th Bohr orbit	20.4×10^{-19}
3rd Bohr orbit	19.4×10^{-19}
2nd Bohr orbit	16.4×10^{-19}
1st Bohr orbit	0

Example #3, violet:

$$f = .75 \times 10^{15} \text{ cycles/sec.}$$

$$\Delta E = hf = 6.6 \times 10^{-34} \frac{\text{joule-sec}}{\text{cycle}} \left(.75 \times 10^{15} \frac{\text{cycles}}{\text{sec}}\right),$$

$$= 4.9 \times 10^{-19} \text{ joules.}$$

D. Matching Energy Level Differences to Energy Levels Predicted by Bohr's Theory

Bohr's theory predicts "allowed" energy levels for the electron in the hydrogen atom, as shown in Figure 2.16.

Match up the energy level differences for the lines you observed to see if there are any two levels on the theoretical diagram that could have caused the radiation you saw.

Example #4, violet:

$$4.9 \times 10^{-19} \text{ joules.}$$

If an electron jumped from the 6th to the 2nd Bohr orbit, it would have lost 4.8×10^{-19} joules, close to our value of 4.9×10^{-19} joules. If the lines match up fairly well, we can be confident that Bohr's theory works for the hydrogen atom.

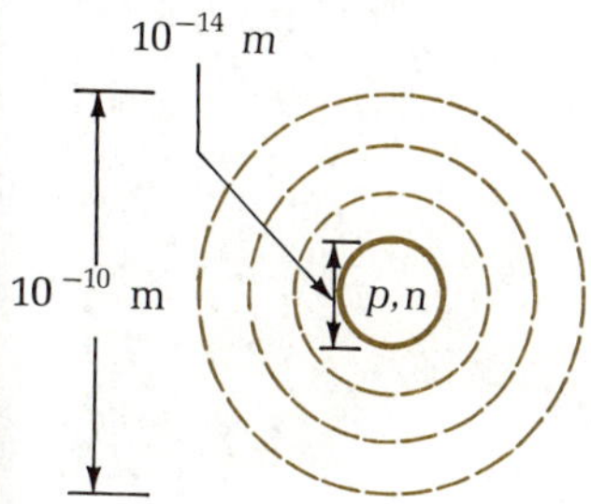

Figure 2.17 *Bohr model of the atom*

Bohr's theory was extended to other atoms besides the hydrogen atom by simply presuming that each electron has a number of definite or "allowed" orbits and, therefore, "allowed" energy levels. When an electron absorbs energy, it jumps to a higher energy level, and when the electron jumps back down, it gives off the energy in the form of radiation according to $\Delta E = hf$. The Bohr model is illustrated in Figure 2.17.

2.7

The Most Modern Model of the Atom— The Cloud Model

The Bohr model of the atom was just great, except for *one little flaw—it only worked for the hydrogen atom*. Energy level diagrams were drawn up for the helium atom, and when the radiation from helium was observed and translated into energy level differences, as you did in Experiment 2.2, the experiment just didn't match the theory.

A new theory was developed by Schrodinger, Heisenberg, Dirac, Born, and others. This theory is much more complex mathematically than Bohr's theory, and is called *quantum mechanics*. This new theory retains the idea of definite energy levels, but abandons the idea of definite orbits for the electrons as they circulate around the nucleus. Instead of fixed orbits, quantum mechanics gives a *probability* of finding the electron at any given spot.

Let's look at an analogy to help us visualize this idea of knowing only the probable rather than the actual location of something. Suppose you have an active white dog on a leash. The leash is fastened to a flagpole on a dark background. Let's say you are interested in the dog's movements, but your presence would excite the dog; so you think of a way to observe the dog without disturbing him. You mount a camera on top of the flagpole, put an automatic timer and shutter on it, set it, and walk away. This whole setup is illustrated in Figure 2.18.

Because the background is so dark, the shutter stays open a long time before it lets in enough light to develop the film. The dog is, of course, oblivious to the camera taking his picture and so he goes about his normal routine of pacing, sniffing, burying bones, and so forth. Because the shutter is open so long, what does the film look like when it is developed? A blur. You can't tell exactly where the dog is, because the dog is in many different places during the time the shutter is open. The predictions of the theory of quantum mechanics about the location of the electrons within the atom are similar to the fuzzy picture of the moving dog.

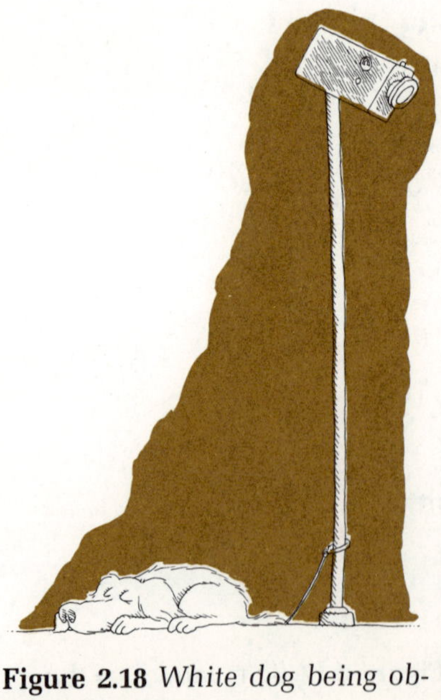

Figure 2.18 *White dog being observed by camera*

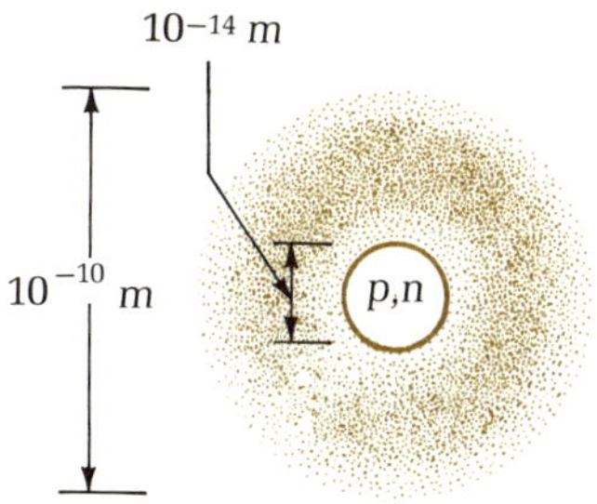

Figure 2.19 *The most modern model of the atom—the cloud model*

Quantum mechanics doesn't tell where the electron *is;* it tells what the probability is of finding electrons at any one place.

If one carried the analogy further by demanding to know where the dog is at one particular instant, and by setting off a flashbulb to make sure there is enough light, the dog's location would certainly be revealed. However, the dog would not resume his normal behavior pattern for some time, because he would be disturbed by the bright light and would undoubtedly set up quite a howl. Similarly, the location of the electron can be determined, but it is determined at the cost of disturbing the electron from its normal path, just as the dog was disturbed. The impossibility of knowing both the position and momentum of the electron is expressed in the *Heisenberg Uncertainty Principle,* which appears to be fundamental to the theory of quantum mechanics.

The theory of quantum mechanics certainly yields a strange picture of the atom, shown in Figure 2.19.

Because the electrons' positions or orbits are not precisely known, the atom is shown with a series of dots which are possible electron positions. Consequently, this model of the atom is referred to as the *cloud model.* (My lab technician refers to it as the foggy meatball model, with the fog representing the electron cloud and the meatball being the nucleus.) Electrons, equal in number to the protons, have definite energies, but their orbits cannot be determined. When electrons jump from one energy level to another lower energy level, they give off radiation according to $\Delta E = hf$.

To some people, this model of the atom seems intellectually dissatisfying. They find it difficult to swallow the notion that the position of the electron *cannot* be known. Physical scientists should be able to know everything; shouldn't they? Few scientists have made such a claim since the disastrous predictions made in the mid 1800's, when some prominent scientists said they doubted anything new would ever be discovered. Scientists learned a lesson from those predictions, and that lesson is never take your hypotheses *too* seriously.

We have seen man's understanding of the atom grow, and his hypothetical model of the atom change as a result of new experimental facts. Each model fit the experiments up to the point of its introduction, and each model was modified by the discovery of *new* facts. The *cloud model* fits experimental data very well at this point in time. Does this mean it is the last model of an atom? Probably not. If new data cannot be explained by the cloud model, someone will have to formulate a new model to explain the data, and so it goes.

Summary

Atoms are at the root of all matter; this chapter looks at how man's understanding of the atom has developed from the first crude model of Democritus to the cloud model.

Democritus envisioned atoms as tiny, indivisible, indestructible spheres. Since Democritus did no experiments, his model was entirely hypothetical.

After the scientific revolution of Galileo's time, J. J. Thomson—while he was experimenting with cathode rays—discovered that small negative charges were contained in atoms. Thomson named these charges *electrons,* and deduced that the atom must also contain enough positive charge to balance the negative charge of the electrons and leave the entire atom electrically neutral. Thomson theorized that the positive charge was "smeared" uniformly throughout the atom and the electrons were "stuck" within this positive charge, like raisins in a plum pudding.

Rutherford experimented with one of the particles given off by radioactive substances—α particles. Although Rutherford set out to confirm Thomson's plum pudding model of the atom, his experimental results indicated that the positive charge was concentrated into a tiny volume he called the *nucleus.* He theorized that the positive charge was carried by a particle called the *proton,* which has a positive charge equal in magnitude to the electron's charge and a mass almost 2000 times that of the electron, and that the electrons orbit the proton-containing nucleus like planets orbiting the sun.

Rutherford's model is refined to include a new particle—the *neutron*—in the nucleus, because chemists who compiled the periodic table of the elements noticed discrepancies in the mass of the atoms, implying that all protons didn't have the same mass. It was hypothesized, and later confirmed, that a neutral particle having approximately the same mass as the proton was also contained in the nucleus.

Next, the pattern of the electrons as they orbit the nucleus was investigated experimentally by forming ions. Electrons were taken from or added to atoms, and the energy necessary to accomplish this ionization was recorded. The results of these experiments led scientists to conclude that electrons are arranged in distinct energy levels, and hence distinct orbits, as they circulate around the nucleus. They further concluded that there are maximum numbers of electrons in each level, and that ions are formed when electrons are taken on or given off to form "filled" energy levels.

Although the energy level model explained ions rather nicely, it did not do anything to explain why atoms radiate light, such as the neon sign, under certain conditions. Borrowing the idea that energy comes in discrete amounts called *quanta* from another problem in physics, Bohr suggested that the electron in the hydrogen atom has several "allowed" orbits, and that radiation is given off when the electron jumps from one of the higher allowed orbits to a lower orbit. Experiments tend to verify that the energies predicted by Bohr's theory correspond rather nicely to the radiation observed. Extending this concept to other atoms implies that each electron has not only its normal energy level and orbit, but a number of "allowed" higher energy orbits, and radiation is given off when the electron jumps from a higher energy allowed orbit to a lower one.

Unfortunately, the predictions of Bohr's theory do not work for atoms other than hydrogen. A more complicated theory called *quantum mechanics* replaces the idea of fixed orbits with probability distributions for the location of the electrons, but retains the idea of fixed energy levels for the electrons. The model of the atom constructed on the basis of quantum mechanics is called the *cloud model*.

Questions

This may be used like a programmed text by referring to the indicated questions while reading the text.

2.1

1. Explain the concept of a model.
2. What was the first model of an atom, formulated by Democritus several hundred years B.C.?
3. For what reasons did Democritus postulate such a model?

2.2

4. What developments in experimental apparatus led to the discovery of the electron?
5. Show how Thomson used the "scientific method" in the discovery of electrons.
6. Show how the existence of a positive charge in the atom may be arrived at by deductive reasoning.
7. What was Thomson's plum pudding model of the atom?

2.3

8. Why is it so difficult to probe into the atom to find out how things are arranged?
9. What was the point of the black box experiment?
10. How could you explore a haystack with a cannon and cannonballs?
11. Explain how the early scientists exploring radioactivity traded one problem for three.
12. Why were the α particles used as atomic cannonballs?
13. Why were Marsden, Geiger, and Rutherford surprised at the result of Marsden's experiment?
14. The ratio of the mass of the proton to the mass of an electron is almost 2000 to 1. Find examples of ordinary objects that have a similar mass ratio (e.g., a compact car and a pound of butter).
15. What was Rutherford's solar system model of the atom?

2.4

16. On what basis did Dalton organize his list of the elements?
17. On what basis did Mendeleev suggest a new arrangement of the elements?
18. How was Mendeleev's scheme justified by later events?
19. Why the big fuss to make sure there are no two elements with the same symbol?
20. Distinguish between the atomic number and the atomic mass.
21. What were the two main difficulties in using the Rutherford electron-and-proton-only model in explaining some of the numbers in the periodic table?
22. How does the neutron answer the difficulties raised in question 21 above?
23. What are isotopes?
24. How many electrons, protons, and neutrons are in an

atom of sodium, $^{23}_{11}Na$? (11 electrons, 11 protons, 12 neutrons)

25. How many electrons, protons, and neutrons are in an atom of nitrogen, $^{14}_{7}N$? (7 electrons, 7 protons, 7 neutrons)
26. Describe Rutherford's model of the atom, which was refined by the addition of the neutron.

2.5

27. What is ionization?
28. What curious features are seen in the graph that shows the amount of energy necessary to singly ionize each atom?
29. How does the complete ionization of some atoms lead to the energy level concept?
30. Explain how the "filled" energy level concept enables one to calculate the number of electrons in an ion.
31. How many electrons, protons, and neutrons are there in the common ion of sodium, $^{23}_{11}Na$? (10 electrons, 11 protons, 12 neutrons, therefore +1 charge)
32. How many electrons, protons, and neutrons are there in the common ion of fluorine, $^{19}_{9}F$? (10 electrons, 9 protons, 10 neutrons, therefore −1 charge)
33. Use the periodic table and the "filled" energy level concept to determine the chemical formula for the ratio of atoms when magnesium, $^{24}_{12}Mg$, and fluorine, $^{19}_{9}F$, combine. (MgF_2)
34. Describe the energy level model of the atom.
35. What property of atoms cannot be explained by the energy level model?

2.6

36. Give an example of radiation of light from atoms.
37. What phenomenon was being explained by Rayleigh and Jeans around 1900?
38. What went wrong with Rayleigh and Jeans' explanation of radiation from heated bodies?
39. How did Max Planck suggest that the assumptions of Rayleigh and Jeans be modified to avert the ultraviolet catastrophe?
40. Give some examples from everyday life to illustrate the difference between discrete and continuous.
41. Explain how Bohr applied Planck's hypothesis to the hydrogen atom.
42. Using the Planck relationship, $\Delta E = hf$, what is the

difference in energy levels which would result in the radiation of light of frequency 5×10^{14} cycles/sec? (3.3×10^{-19} joules)

43. Using the Planck relationship, $\Delta E = hf$, what is the difference in energy levels which would result in the radiation of light of frequency 7×10^{14} cycles/sec? (4.6×10^{-19} joules)
44. Think of an analogy to demonstrate the concept of "allowed" energy levels for the electron in the hydrogen atom.
45. How was Bohr's theory extended to atoms other than the hydrogen atom?
46. Describe the Bohr model of the atom.

2.7

47. What was wrong with the Bohr model of the atom?
48. What portion of the Bohr model was abandoned because of quantum mechanics?
49. Describe the white-dog analogy and show how it is analogous to the electron circulating around the nucleus.
50. What happens if the electron is "illuminated" so that its position may be determined?
51. Describe the cloud model of the atom.
52. Explain why many scientists are not too disturbed by the fact that the cloud model predicts that the exact path of the electron cannot be known.

Quotes to Ponder

1. "It was quite the most incredible event that has ever happened to me in my life. It was almost as incredible as if you had fired a 15-inch shell at a piece of tissue paper and it came back and hit you." (Lord Rutherford)

2. "Students in my chemistry class have come to appreciate my puns. In fact, one girl went so far as to exclaim in class one day: 'Mr. C., you're just a big boron!' Another student, a very conscientious sort, always reminded me when I strayed too far from my lesson plan: 'Sir, isn't it time you returned to the bismuth at hand?'

Not all students, however, appreciate the fine art of punmanship. One lad even suggested, quite seriously, that I was either a cadium or victim of dysprosium. That really shook me for awhile. But my wife, good samarium that she is, quickly reassured me that, other than my weakness for puns, I was quite normal (or was it molar?).

Then and there, I made a firm promethium: Never again would I pun. But the rhodium to hell is paved with good intentions. Within a week, I was tossing puns again—I just couldn't get iridium." (*Chemistry,* 44:4, April, 1971)

3. "Scientists working with a two-mile-long atom smasher suspect they have discovered particles even tinier than protons or neutrons, the smallest known components of an atom. They call the particles 'partons.'

The announcement was made Tuesday by a team of scientists at the Stanford Linear Accelerator Center.

Dr. Wolfgang K. H. Panofsky, director of the center, and Dr. Henry W. Kendall of Massachusetts Institute of Technology, also reported the results in the June issue of *Scientific American* magazine.

It has been thought that protons and neutrons, which form the core of an atom's nucleus, could not be divided.

But bombardment of these particles by streams of electrons of up to 21 billion electron volts has suggested that both protons and neutrons have internal structural parts, called 'partons,' Panofsky said.

'Knowledge of the internal structures of the protons and neutrons may provide the key to understanding the strong force that holds the atomic nucleus together and endows the universe with its stability,' he said." (Associated Press News Release, 1971)

CHAPTER THREE

HOW FAR APART ARE THE ATOMS IN SOLID MATTER?

Solid matter looks like a good place to begin studying our environment. We ourselves *are* solid matter, and we are certainly surrounded by a wealth of different kinds of solids. An in-depth study of solid matter, however, presents some difficult problems for scientists to ponder. The first problem posed sounds innocuous enough: Since all matter consists of atoms in one form or another, how far apart are the atoms which make up solid matter? It would be nice if we could just take a close look at a solid and be able to distinguish one atom from another. Unfortunately, neither magnifying glasses nor microscopes will do the trick; those darn atoms are just too small for us to measure by ordinary means. We'll take a close look at, and make some hypotheses about, the fundamental structure of crystals, one kind of solid matter. Then, we'll take what will apear to be a side trip into light and other kinds of electromagnetic radiation. Next, we'll use one particular kind of electromagnetic radiation—x-rays—to analyze crystals and determine their basic structure, and thus finally arrive at an answer to the question, How far apart are the atoms in solid matter?

3.1

One Particular Form of Solids—Crystals

We could begin analyzing almost any kind of solid matter, but there are good reasons for starting with crystals. They are not only beautiful, they are the simplest form of solid matter to analyze because of their regularity. The first thing to do is to make some crystals to examine.

EXPERIMENT 3.1
HOME GROWN CRYSTALS

A. *You will be given a jar containing some powder, which will eventually yield some form of crystal(s).*

1. *Observe the contents of the jar carefully, using a*

magnifying glass. Record a description of the contents.

2. Obtain enough distilled water (or as pure water as you can get) to fill the jar, but don't put the water into the jar yet.

3. Using a pot or teakettle, heat the water until it is almost boiling.

4. Pour the hot water into the jar containing the powder to about one centimeter from the top. Screw the lid on tightly and shake the jar and contents vigorously until you see no more undissolved powder. Then shake for five minutes more. (CAUTION: The fluid may be poisonous; so keep it out of the reach of small children.)

5. Take the lid off and place the jar and its contents in a place where it will not be disturbed for several weeks.

6. Without touching the jar, observe it every day or so. For the first couple of weeks, nothing should happen besides a decrease in the fluid level.

7. When the fluid level gets near the bottom of the jar, don't despair; your crystal is probably just forming. As soon as you see a bit of crystal poking its head above the fluid, stop the experiment by putting the lid back on the jar. Bring the jar, lid, crystal, and remaining liquid to the next lab session to be analyzed. Make sure you record a description of the crystal(s) that formed.

8. The results of this experiment will be explained in Chapter 5, when we develop a conceptual model for solids.

B. Salt and Sugar Crystals

To make salt crystals from ordinary table salt, start by dissolving as much salt as you can in about a third of a glass of (distilled, if possible) water. Keep adding salt and stirring vigorously until no more salt will dissolve. You now have a glass containing water, salt, and cornstarch. (The cornstarch was added by the manufacturer to keep the salt from sticking together in wet weather. When it rains, it pours, you know!) Cornstarch will not dissolve in water, and it will interfere with the crystal formation process. Therefore, let the cornstarch settle to the bottom of the glass for a couple of hours or so, and pour off (chemists call this decanting) the clear solution containing the dissolved salt. (Red wine and old-fashioned iced tea drinkers have to contend with this sediment problem constantly.) Put the salt solution in some place where it will not be disturbed, and observe it periodically for a few days.

Sugar crystals may be formed by dissolving half a cup

of sugar in a quarter cup of hot (preferably distilled) water. When the water has cooled and evaporated for a day or so, sprinkle in some sugar granules and see if they grow into crystals.

Now that we've started some crystals growing, let's have a look at some which are already grown to discover some of their properties.

EXPERIMENT 3.2
SALOL CRYSTALS

You will be supplied with a glass slide on which there are two blobs of a substance called salol (pronounced say′ loll).

First of all, try to describe the substance as completely as you can. With your fingernail, chip away a small piece of the small blob and examine it closely with a magnifying glass.

Some descriptions may be more eloquent than others, but yours should include the facts that salol is clear to whitish in color, crystalline in form, and shaped similar to a diamond (as in clubs, diamonds, hearts, and spades).

Next, using the matches provided, melt the large blob by holding the match near the bottom of the slide (see Figure 3.1). Don't hold the match too close so that you blacken the slide, but make sure the entire blob is melted. Use more than one match if necessary.

Observe the liquid, and describe it in your notes.

After the liquid cools back down to room temperature, you would expect the crystals to reform. After all, this stuff was a solid at room temperature when the experiment began, and, for that matter, the little blob is still solid.

Hang on to your intuition and your reasoning powers! It doesn't return to the solid form all by itself. (If it does, you probably did something wrong. Reheat it and try again.)

Poke it with your pencil or pen and put the matchstick into it; it still remains liquid.

Next, shove some of the broken-off pieces from the little blob into the melt, and watch what happens with the magnifying glass. Beautiful! It recrystallizes before your very eyes.

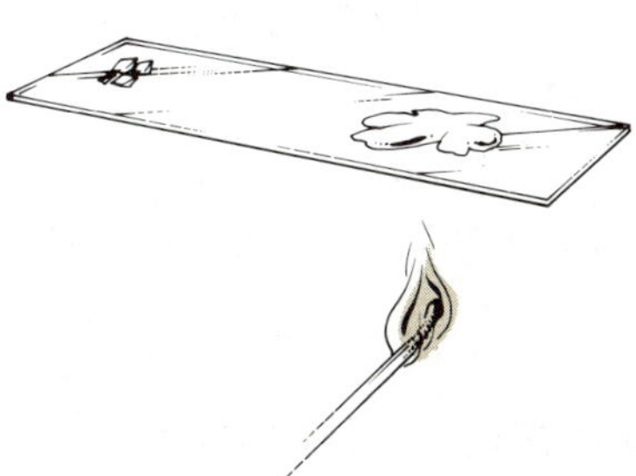

Figure 3.1 Melting salol crystals

This experiment is not just a magic trick to amaze and mystify you. (Although a little magic doesn't hurt.) Besides giving you a close look at a crystalline solid, it allows you *to make an observation that must be explained by any hypothetical model that we come up with. Our theoretical model for solids must offer a plausible explanation for this supercooling phenomenon at the atomic level. We'll try to put this together in Chapter 5.*

Before we really begin discussing crystals in earnest, you've got to do one more experiment to become even more familiar with them.

EXPERIMENT 3.3

CRYSTAL CLEAVING

Crystals don't exactly come apart like other solids. You will be given some crystals, a single-edge razor blade, and a hammer. The idea is a fairly simple one; by observing the natural faces on the crystals you are given, try to figure out where to strike the crystal with the razor blade to make it come apart with smooth faces, or facets. This process is called cleaving, and is used in the preparation of many gem stones, notably diamonds. (Undoubtedly, many of you have seen an old movie or television commercial where some sweating expert cleaves a diamond perfectly under some kind of adverse condition.)

The cleaving technique should be a sharp rap, similar to the way you would crack an egg on the edge of a skillet, as opposed to a rock-crushing blow that you might use on a stubborn walnut. Safety glasses are recommended.

Cleave the crystals into as small pieces as you can and examine them with the magnifying glass. Sketch the crystals for your notes. As you cleave these crystals, try to envision what sort of fundamental arrangement of atoms would allow crystals to cleave like this.

Now that you've seen some crystals "up close," you have undoubtedly been impressed with several characteristic features of crystals:

1. Crystals have shiny, extremely flat faces.
2. All crystals of the same material look the same; that is, the angles between the faces appear to be the same for crystals formed from the same material. See Figure 3.2.

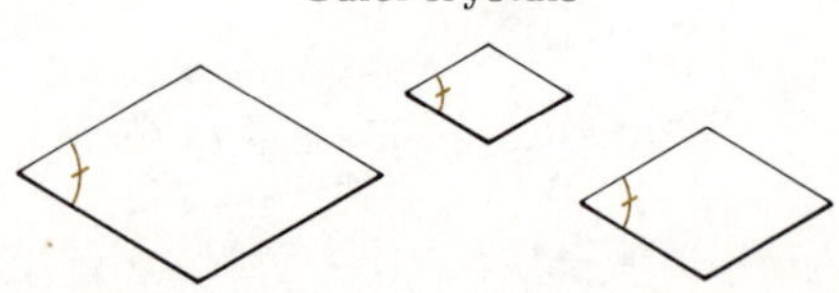

Figure 3.2 *Sketches of salol crystals*

Figure 3.3 *Calcite crystal*

3. Some crystals cleave; that is, they come apart, forming smooth faces when they are struck with a razor blade parallel to existing smooth faces.

The term *crystal* is derived from the Greek *crystallos,* applied to quartz by the geographer Strabo about 2000 years ago. Characteristically, crystals were discussed and sought after by alchemists, but few quantitative experiments were performed until after the time of Galileo. The first experiments were done with a fairly crude device that measured the angles between faces of crystals. These measurements confirmed the notion that the angle between faces is the same for all crystals of the same material.

The constancy of the angles between faces and the cleaving capabilities of crystals caused a number of scientists to suspect that there must be some regular, symmetric structure underlying crystals. René Hauy (pronounced Howie), a scientist of the late 1700's, accidentally dropped a calcite crystal and it cleaved naturally, exhibiting a beautiful, smooth, shiny face (Figure 3.3).

"The prism had a single fracture along one of the edges of the base," he wrote later, "by which it had been attached to the rest of the group. Instead of placing it in the collection which I was then making, I tried to divide it in other directions . . ." He found that there were three directions in the crystal along which he could break it and get a bright plane surface like the first one.

When Hauy observed that the angular relations between the cleavage faces were the same in every fragment, he erroneously concluded that he could break the crystal into the small pieces which were the ultimate building blocks of which it was constructed: ". . . and I succeeded after several trials," he wrote, "in extracting its rhomboid nucleus." In spite of this error, he did have the concept of a repeat unit, and he showed that he could reconstruct the original sharply pointed crystals, with the proper interfacial angles, by stacking up the little cleavage blocks which he called "molecules integrantes."*

Hauy suspected—as you perhaps did, as you cleaved the crystals in Experiment 3.3—that there are some *fundamental building blocks* from which crystals are formed, just like a brick wall is made up of a regular arrangement of

---

*From Elizabeth A. Wood, *Crystals and Light: An Introduction to Optical Crystallography* (New York: Van Nostrand Reinhold, 1964).

bricks. This conceptual model of the fundamental structure of crystals explains the different shapes of crystals by saying that crystals are made of differently shaped building blocks.

For example, suppose we use a *cube* as a fundamental building block. It's not too difficult to envision cubic building blocks assembled to form a large cubic crystal. Common table salt (NaCl) is a cubic crystal; so we might theorize that NaCl is made of small cubic building blocks (see Figure 3.4). The NaCl crystal would have flat faces because the building blocks have flat faces. The angles between faces would be constant because the building blocks would be stacked with constant face angles, and the crystal would cleave between the building blocks.

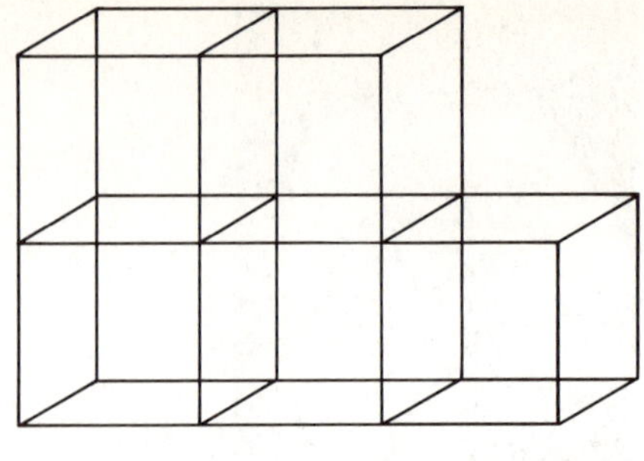

Figure 3.4 *Cubic building blocks assembled to form a crystal like NaCl*

In the mid-1800's, several scientists extended this building block notion using geometrical methods. They postulated the existence of many different shapes of fundamental building blocks which form all the known crystals (see Figure 3.5).

Wait a minute! We're progressing so smoothly with these crystal fundamental building blocks and all their fascinating geometric shapes, but we're missing the whole point of what we discussed in Chapter 2.

Where do the atoms go in these fundamental building blocks?

We know that atoms are somewhat "fuzzy" because of their electron clouds, so we could hardly expect the atoms themselves to assume such nice, regular, geometric shapes. Let's expand our model to include atoms by presuming that atoms could be located at corners, along edges, in the center of a face, or in the center of the whole structure (body center), as illustrated in Figure 3.6.

Now the question of how far apart are the atoms can be asked in terms of the relative size of the atoms and the basic

Figure 3.5 *Some of the geometric shapes of crystal fundamental building blocks*

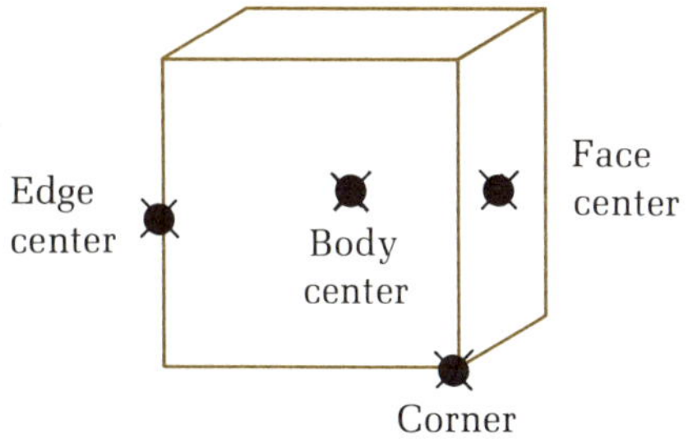

Figure 3.6 *Possible atom locations in a cubic crystal building block*

building block structure. The possibilities may range from having the atoms fill the structure to having the atoms occupy only small points in the structure (see Figure 3.7).

Regardless of the atom size, since the building block structure has these atoms in it, we have to ask the question:

How can crystal building blocks contain all these "lumpy" atoms with their electron clouds of indefinite shape and still "stack up" in such a way that the faces appear to be smooth and flat, even to the point of reflecting light like a mirror?

Why don't we see a confused reflection pattern because of all the "bumps" made by the atoms?

Before we can answer the questions about crystal building block sizes and reflections from atoms, we'll have to take a substantial sidetrip into light and other forms of electromagnetic radiation. Don't give up; it will all fit together eventually.

3.2

Light and Other Electromagnetic Radiation

Similar to the development of the conceptual model of atoms, discussed in Chapter 2, physical scientists developed conceptual models of light in an attempt to understand its nature. The two main theories developed are the *particle theory* and the *wave theory*.

The *particle theory* says that light consists of a stream

Figure 3.7 *Possible atom-building block size combinations*

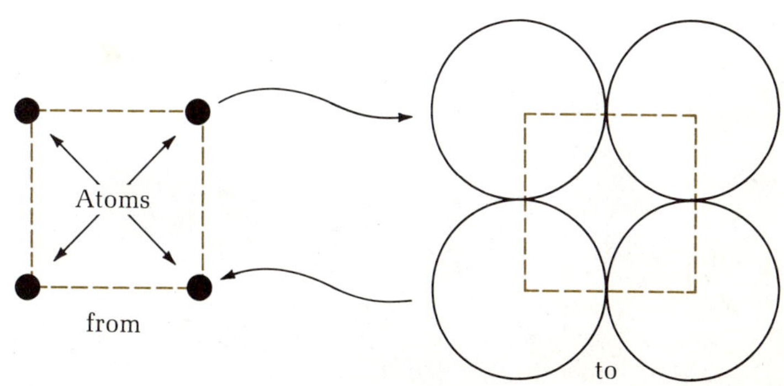

of tiny particles which move in straight lines at very high speed (see Figure 3.8).

While it isn't too difficult to envision a stream of light "pellets" zipping along, the *wave theory of light* will be a little more difficult to explain; so let's start with water waves. We've undoubtedly had some experience with water waves, and we can generalize from them to light and other electromagnetic waves.

Figure 3.8 *"Particles" of light*

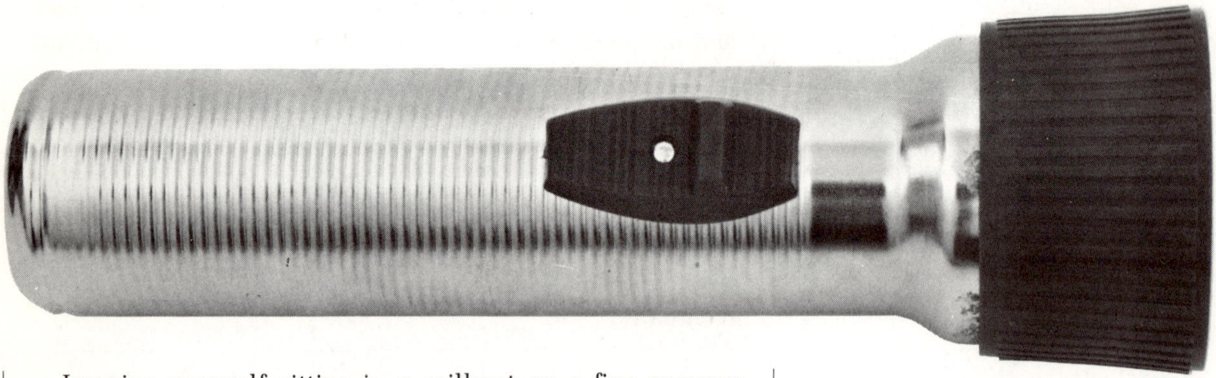

Imagine yourself sitting in a sailboat on a fine summer day. There was just enough wind to get you out into the middle of the lake, then the wind died. It is such a fine day that you are perfectly content to just sit there and let the world go by. Suddenly you hear an approaching motor, and a speedboat zooms by. You don't notice anything immediately, but eventually the boat begins to bob up and down. If we cut a cross section through the lake, we'd see the wave pattern in Figure 3.9.

The wave is a disturbance of the water's surface, and this disturbance moves *across* the surface while the surface moves only up and down. A water wave is said to be *transverse* because the wave moves perpendicular to the motion of the medium. (Not all waves are transverse. Sound waves alternately compress and expand the medium through which they travel, and are said to be *longitudinal*.) In Figure 3.10, let's take a closer look at one of the waves.

The "top" of the wave is called a *crest*, while the "bottom" is called a *trough*. (Sounds like something horses and pigs

Figure 3.9 *Water waves*

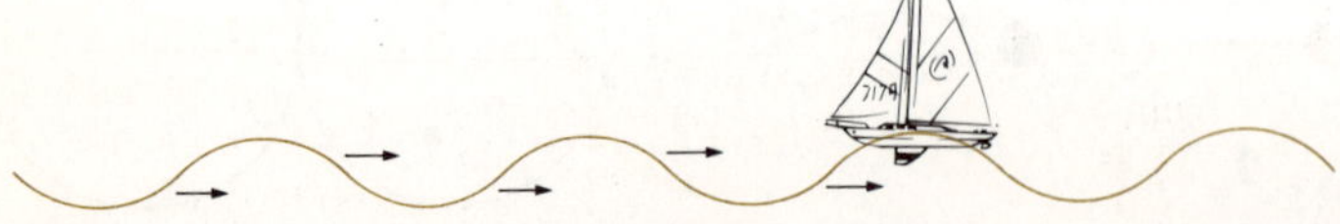

drink from.) The point where the wave crosses the line which represents the undisturbed surface is called a *node*, and the height of the crest above the node is called the *amplitude*.

The distance from crest to crest (or from trough to trough) is called the *wavelength*. It is represented by the Greek letter lambda, λ.

Returning to your becalmed sailboat, suppose you try to figure out how fast the waves, caused by passing motorboats, are moving. Looking at your trusty wristwatch (waterproof, of course) you begin timing the up-and-down motion. You find it takes t seconds to go from one crest to the next. Since crests are separated in distance by one wavelength, λ, the wave speed must be

$$\text{Wave speed} = \frac{\text{Distance traveled by wave}}{\text{Time it took to travel that distance}},$$

$$v = \frac{\lambda}{t}.$$

The time t is called the *period* of the wave, since it is the period of time necessary to complete one full cycle. It has dimensions of seconds/cycle. A quantity that is even more useful is the reciprocal of the period, called the *frequency*, f.

$$f = \frac{1}{t} \text{ (cycles/sec)}.$$

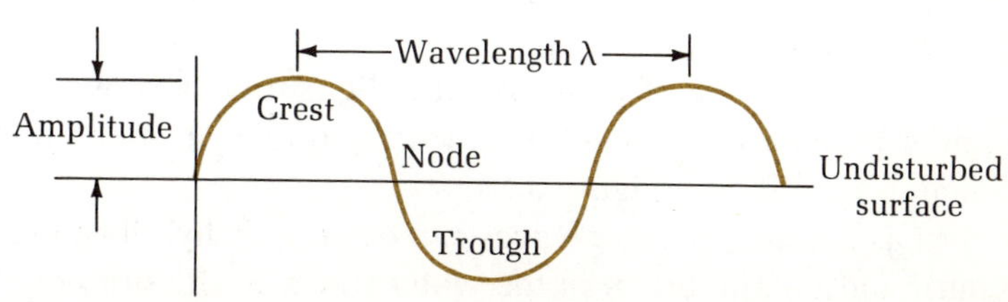

Figure 3.10 *Wave close-up*

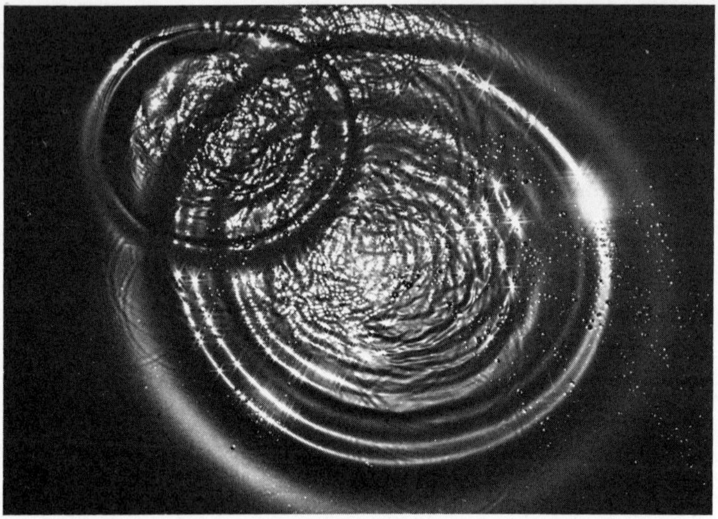

Figure 3.11 *Light "waves" are similar to water waves*

While the period tells how many seconds of time are necessary to complete one cycle, the frequency tells how many cycles are completed in one second. (Better read that a couple of times to make sure you've got it.) The equation for wave speed becomes

$$v = f\lambda.$$

For example, suppose the waves appear to be about 2 meters between crests, and your watch (and perhaps your stomach) tells you that the crests are coming at the rate of 3 per second, then the wave velocity must be

$$v = \left(3\ \frac{\text{cycles}}{\text{sec}}\right)\left(2\ \frac{\text{meters}}{\text{cycle}}\right) = 6\ \frac{\text{meters}}{\text{sec}}.$$

As another example, suppose you know the wave speed to be 18 meters/second, and the wavelength is 3 meters/cycle. What is the frequency?

$$f = \frac{v}{\lambda} = \frac{18\ \text{m/sec}}{3\ \text{m/cycle}} = 6\ \frac{\text{cycles}}{\text{sec}}.$$

Now for the big question: How are water waves and light waves similar?

The wave theory of light says that light travels in waves similar to those that would result from dropping a stone into a smooth pond. See Figure 3.11.

Just as a series of water waves can be created by vibrating your hand up and down in the water (try it in the tub with

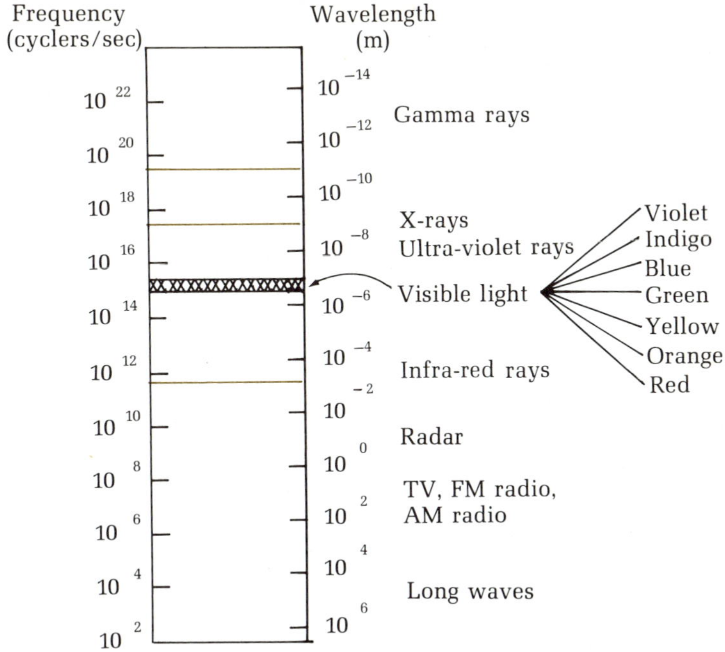

Figure 3.12 *Electromagnetic spectrum*

your rubber ducky), light waves are thought to be created by a vibrating electric charge, such as an electron.

Because light waves are thought to be caused by vibrating electric charges, these waves are called *electromagnetic*. A whole *spectrum* of waves of different frequency and wavelength exist. All electromagnetic waves do have one property in common—they all travel at the same speed, $c = 3 \times 10^8$ m/sec (in vacuum), referred to as the *speed of light*. Since electromagnetic waves are similar to water waves, their wavelength, frequency, and speed must follow the same relationship.

$$f\lambda = c = \text{speed of light} = 3 \times 10^8 \text{ m/sec.}$$

Depending on the frequency of the vibrating electric charge that is the source of the waves, the waves may be gamma rays (remember the radioactivity from Chapter 2); x-rays (used frequently by doctors and dentists); ultraviolet rays (which cause suntan and sunburn); visible light; infrared rays (used for heating food); radar (for keeping track of airplanes and weather); and radio and television waves (to entertain us). The frequency and wavelength of these regions are sketched roughly in Figure 3.12 above. Note the small portion of the spectrum occupied by visible light.

Since all electromagnetic waves travel at the same speed,

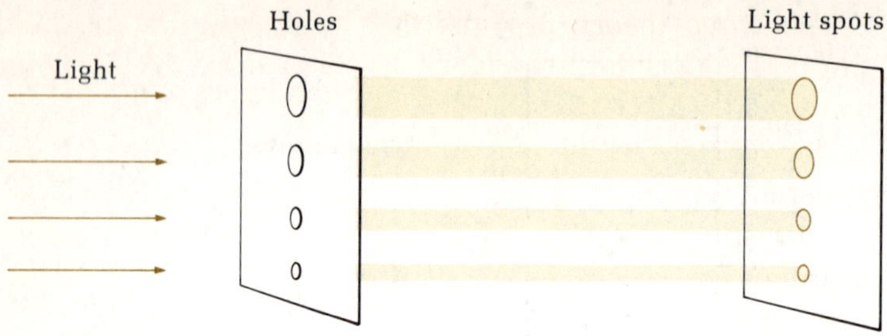

Figure 3.13 *Particle theory prediction for experiment*

c, it is possible to find the wavelength of an electromagnetic wave, given its frequency (or vice versa), by using the relationship

$$f\lambda = c = 3 \times 10^8 \text{ m/sec}.$$

For example, what is the frequency of a 1 cm radar wave?

$$f = \frac{c}{\lambda} = \frac{3 \times 10^8 \text{ m/sec}}{1 \times 10^{-2} \text{ m/cycle}} = 3 \times 10^{10} \text{ cycles/sec}.$$

What is the wavelength corresponding to a gamma ray of frequency 6×10^{21} cycles/sec?

$$\lambda = \frac{c}{f} = \frac{3 \times 10^8 \text{ m/sec}}{6 \times 10^{21} \text{ cycles/sec}} = 5 \times 10^{-14} \text{ m/cycle}.$$

If we regard the particle theory and the wave theory of light as competitors, we can decide which is the "winner" by having each theory make a prediction of the result of an experiment. Then, we can perform the experiment and see which theory is justified by the experimental result.

For the experiment, let's simply *shine light through a series of holes of various diameters.*

The *particle theory* predicts that the light particles would zip right through the holes and reproduce the hole pattern in light as shown in Figure 3.13.

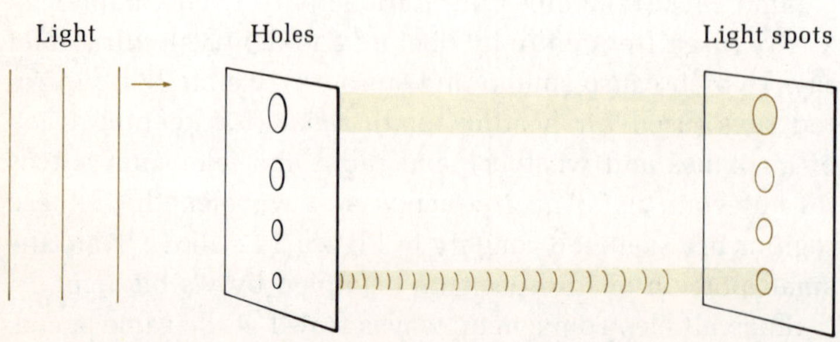

Figure 3.14 *Wave theory prediction for experiment*

The *wave theory* predicts that each point in the hole through which the light is shining would act like a point source of light waves, as shown in Figure 3.14.

Notice that the "spreading" of the waves is much more apparent with the smallest hole than with the largest hole.

This spreading of waves at a small opening is called *diffraction*. Let's do the experiment and see what we'll get for results.

EXPERIMENT 3.4

DIFFRACTION OF LIGHT

It is easier to do an experiment showing diffraction than it is to explain diffraction in words. All you need for this experiment is a bright light, a pin, a pencil, some paper, and a dark room. (Try not to fall asleep until after the experiment.)

The idea is simply to punch a series of holes of various diameters in the paper and shine the light through the holes one at a time, observing the size of the light spot on the wall.

What you should see is illustrated in Figure 3.15.

It appears that the particle theory works for the large holes and the wave theory works for the small holes! Don't let this bother you too much; physicists have had to live with it for years. It seems that light behaves like a stream of particles when the holes are much larger than a wavelength of the light involved, and it behaves like a series of waves when the holes are roughly comparable in size to the wavelength of light involved.

Since we're interested in crystal building blocks, which are undoubtedly quite small, let's pursue the wave theory a bit farther and have a look at the phenomenon called interference.

Figure 3.15 *Diffraction experiment setup*

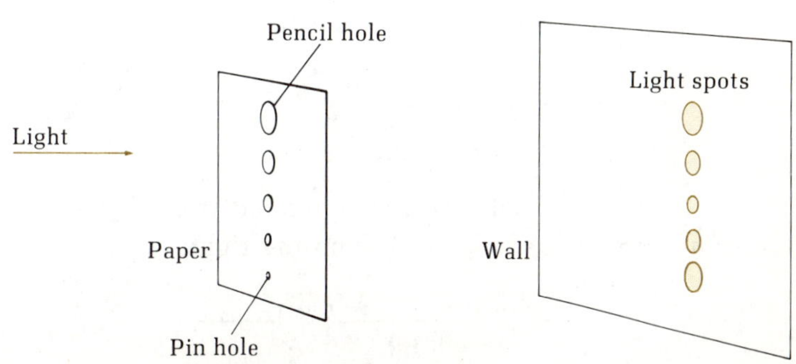

Suppose you shined the light through two narrow openings, called slits, which were very close together. Wave theory says there would be diffraction at both slits, and the light would thus spread out in circular wave forms as shown in Figure 3.16.

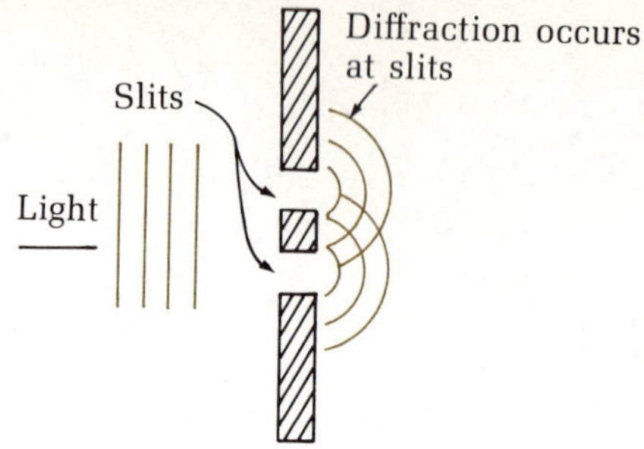

Figure 3.16 *Light diffracting at two slits*

Two light waves would attempt to occupy the same point at the same time; this is called *interference.* (Should sound familiar to you football fans.) We know what happens when two solid objects try to do this, but what about waves?

To dig out the answer to this one, let's look at the way waves come together on a long spring. (You might try this with a long "Slinky.") Start some kind of wave form moving from one end toward the other (Figure 3.17).

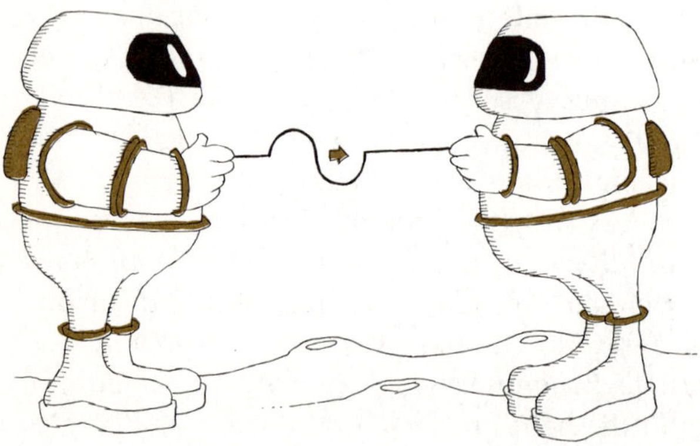

Figure 3.17 *Wave form moving on spring (a–l)*

It moves along nicely enough,

(b)

until it gets to the other end, where it is reflected and goes back in the other direction with crests reflecting as troughs and troughs reflecting as crests.

(c)

Now suppose a single crest is generated from the left at the same time as a single crest from the right:

(d)

What happens at the instant of time when 1 meets 2? Just as you would probably expect, you get a momentary double crest.

(e)

The waves never stop; so just after this double crest is formed, the waves move on through:

(f)

Similarly, a trough meeting a trough yields a double trough momentarily as they move through each other,

(g)

(h)

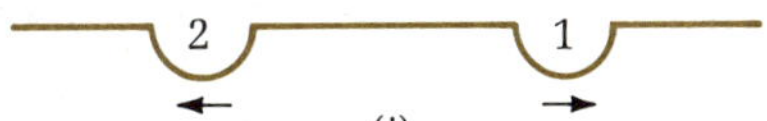

(i)

and a crest and a trough meet to make a momentary node before they pass on.

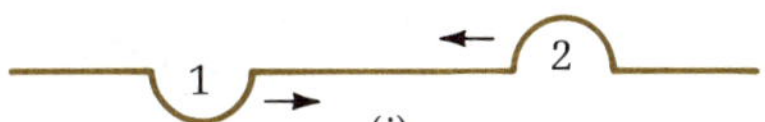

(j)

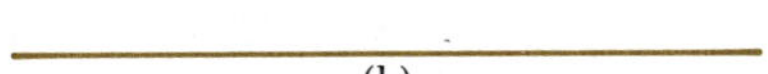

(k)

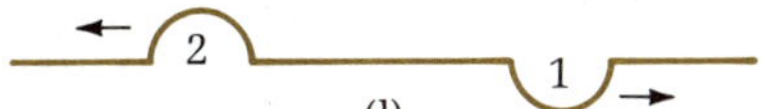

(l)

This process of combining two waves by simply adding the displacement of one to the displacement of the other is called the *principle of superposition*.

You can create some very interesting patterns by generating a series of crests at one end, which then get reflected as troughs. If you generate them at just the right intervals, you can get the crests and troughs to always meet at the same spot on the spring, which makes a permanent node, or spot which doesn't move. These are called *standing wave patterns* and may look like the following illustration. (Try it; it's easier to do than to describe.)

Standing waves

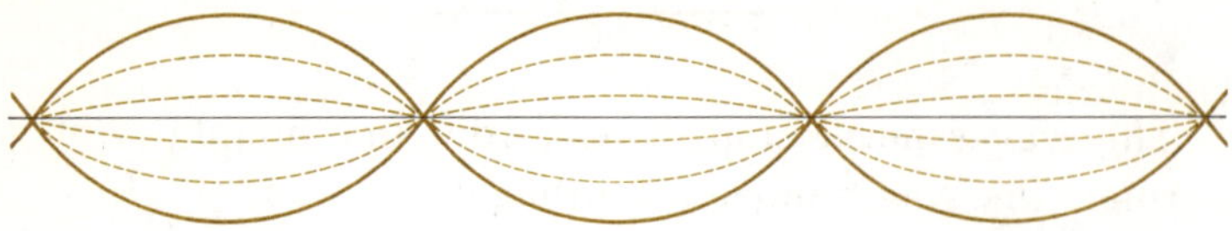

Electromagnetic waves should behave the same way as the waves on a spring. Therefore, we should be able to use the principle of superposition to obtain the resulting wave pattern by simply adding the amplitudes of the interfering waves. That is, two crests will add up to a double crest; two troughs will add up to a double trough; and a crest and a trough will add up to a node. The way we'll be able to detect these "interfered" waves is as follows:

Crests or troughs correspond to bright spots;
nodes correspond to dark spots.

Normally, when you see a light wave, it is coming at you so fast, you can't see the individual bright or dark spots. Instead, you see an "average" somewhere between bright and dark. When the light diffracts through two slits, however, the bright and dark areas are reinforced, making dark and light lines visible, as shown in Figure 3.18.

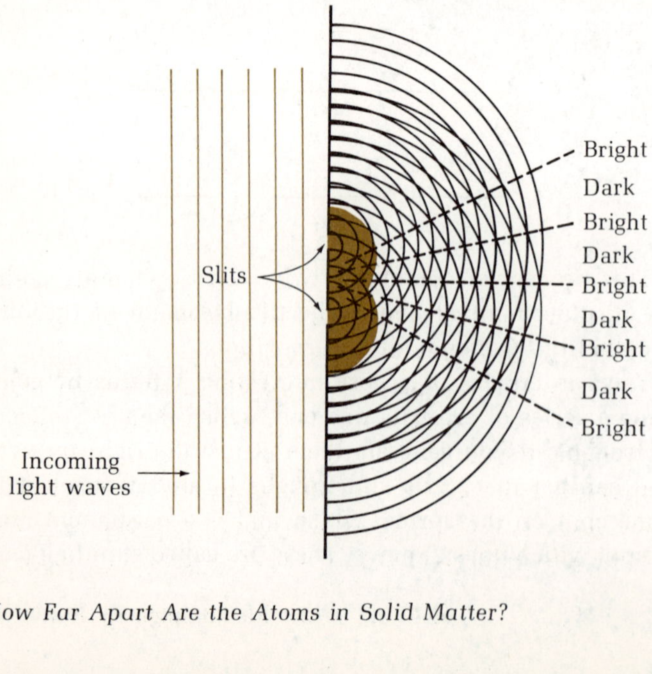

Figure 3.18 *Wave theory prediction for diffraction and interference from a double slit*

At all the spots marked "bright" in Figure 3.18, a crest from one slit is meeting a crest from the other slit, or a trough from one slit is meeting a trough from the other slit. At all the spots marked "dark" in Figure 3.18, a crest from one slit meets a trough from the other slit. The wave theory thus predicts that if a screen is set up, it would show a series of alternating bright and dark lines. Before we do this experiment, let's analyze this situation just a bit farther and come up with a quantitative relationship that we can check out experimentally. First, let's draw a diagram so we can do some geometry (see Figure 3.19).

The bright spot in the center (point B) was formed by light waves diffracting at each slit (s_1 and s_2). Each wave moved an equal distance so that both would arrive at point B at a crest; therefore, crest and crest meet and form a bright spot. The bright spot just to one side of the center (point C) must also be formed by a crest adding to a crest; but now the wave from slit s_1 *must travel farther* because the bright spot is closer to slit s_2. How much farther? Well, since both must end on crests to form a bright spot, the wave must travel a distance equal to the distance from one crest to the next crest. This distance is *one wavelength,* or λ. The light from slit s_1 must travel one crest farther than the light from slit s_2. This distance is one whole wavelength. (By the way, the other bright spots are formed when the light wave travels two whole wavelengths, and so forth, farther from the upper slit.)

Now let's do some geometry to obtain a quantitative relationship that we can use in an experiment. We'll form two similar triangles and equate ratios of their sides.

First, consider the triangle s_1s_2A. This triangle has been

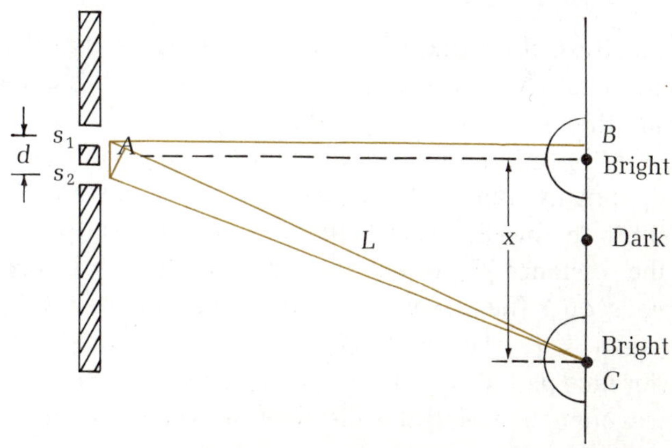

Figure 3.19 *Another view of the double slit apparatus (not drawn to scale)*

formed by drawing a perpendicular from the line s_1C down to the slit s_2. The distance s_1A represents the *additional* distance traveled by the light wave from slit s_1 over that traveled by the light wave from slit s_2. This distance s_1A is one whole wavelength, λ. The side opposite the right angle in this triangle is the distance between slits s_1 and s_2. This distance is given the symbol d. The ratio of these two sides is λ/d.

Now consider the triangle s_1BC. BC is the distance from the middle of the first bright spot away from the center, to the middle of the center bright spot (marked x), plus half the distance between the slits ($d/2$). Triangle s_1BC is similar to triangle s_1s_2A. (If you remember your geometry, you'll recognize that these triangles are similar because they each have two sides perpendicular to each other. If you don't remember your geometry, try constructing these triangles with straight edges until you are convinced that they are similar.) The ratio of the two sides corresponding to the ratio of sides we chose in triangle s_1s_2A is:

$$\frac{x + d/2}{L}.$$

Since s_1s_2A and s_1BC are similar triangles, we can equate the ratios of their corresponding sides to obtain:

$$\frac{\lambda}{d} = \frac{x + d/2}{L}.$$

Let's make this equation simpler by specifying that L will always be so much larger than $d/2$ that we'll ignore $d/2L$. The equation becomes:

$$\frac{\lambda}{d} = \frac{x}{L}$$

Let's analyze this equation just a bit. It states that if light of wavelength λ undergoes diffraction at two slits a distance d apart, they will form an interference pattern. The distance between the middle of the center bright spot and the first bright spot to either side (x) divided by the distance from the slit to the bright spot is equal to the wavelength divided by the distance between slits. Notice that the pattern "spreads out" (larger x) as the distance between slits (d) *decreases*. Let's see what happens to the "spread" of the interference pattern (x/L) for various relationships between the wavelength, λ, and the distance between slits, d.

Case #1: d very much greater than λ. This case corresponds to a large distance between slits, and since λ/d becomes very small, x/L must become small also, and we don't see any interference effects.

Case #2: d greater than λ. As d begins to approach λ, x/L becomes larger, and the interference pattern "spreads out."

Case #3: d less than or equal to λ. The pattern cannot spread out beyond the point where $x/L = 1$, that is, where $\lambda = d$. Thus, if d is less than λ, we would again see no interference patterns.

Since we have this relationship, let's do an experiment to check it out.

EXPERIMENT 3.5

YOUNG'S DOUBLE SLIT EXPERIMENT

The experiment in which light is passed through two very close slits was originally done by Sir Thomas Young, and so it is referred to as Young's double slit experiment. Our goal is to see if the prediction of the wave theory of light may be verified by experimental evidence. Our procedure will be to look at a light source through a double slit, measure the geometric distances called for in the equation, solve for the wavelength of the light, and compare this measured wavelength with the known wavelength for light of that color. If the measured wavelength matches the known wavelength, that means the wave theory prediction was a good one, and we'll be able to use the wave theory to probe solid matter with light.

Procedure

1. *Set up the apparatus as shown in Figure 3.20.*

2. *From a distance of about one meter, look through the slits toward the light source. You should see a series of bright and dark lines of the same color as the light source. Look partly through the slits and partly through the "window" above the slits so that you can see the bright and dark lines and the cardboard pieces placed on the meter stick at the same time.*

3. *Have your partner slide the cardboard pieces along the meter stick until one cardboard piece seems to have its edge along the middle of the center bright line and the other*

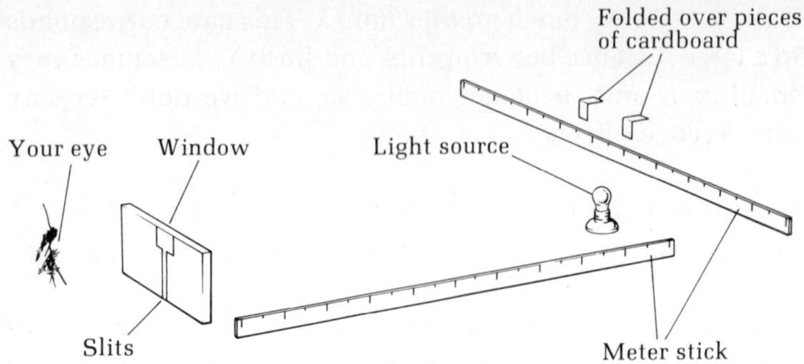

Figure 3.20 *Young's double slit experiment setup*

cardboard piece seems to have its edge along the middle of the bright line next to the center bright line.

4. *Record the distance between the cardboard edges (x).*

5. *Record the distance between the slits and the cardboard edge over the middle of the first bright line next to the center bright line (L). (Refer to Figure 3.19 if you're getting confused.)*

6. *Determine the distance between slits (d). The slits were drawn with two double-edge razor blades next to each other, so you might use a metric micrometer to measure the width of one blade; or else stack up a bunch of them and measure their thickness with a meter stick.*

7. *Using your measured values for x, L, and d, solve the equation $\lambda/d = (x + d/2)/L$ for λ.*

8. *Compare your values for the light wavelength with the following known (measured by other experiments) values.*

Color	Wavelength
Red	6.6×10^{-7} meters
Orange	6.1×10^{-7} meters
Yellow	5.9×10^{-7} meters
Green	5.4×10^{-7} meters
Blue	4.6×10^{-7} meters
Violet	4.0×10^{-7} meters

Your values may vary a bit depending on the exact shade of the colored light and the accuracy of your measurement, but your values should be close to the ones given here for us to have any confidence in the wave theory of light.

Now that we have had some experience with light, let's

extend this knowledge a bit more and try to shed some light (Oops, another bad pun!) on crystals.

Suppose there was a *whole series of slits, rather than just two*. What would the interference pattern look like? As long as the slits are regularly spaced, the interference patterns still look like a series of bright and dark lines. Such things are called *diffraction gratings* and are sometimes used to generate interesting lighting effects.

Suppose that, rather than a whole series of slits which allow light to pass through, there was a *whole series of tiny mirrors which would reflect the light.*

As long as the spacing between mirrors is anywhere between the size of the wavelength and a factor of a few hundred or so larger, we would see an interference pattern just as we did before. If the spacing between mirrors is either much greater or else less than the wavelength of the light waves, we would see no interference patterns, only reflections.

Now we're ready to make a big leap—rather than a regular arrangement of tiny mirrors, how about a regular arrangement of atoms? We suspect that a crystal consists of a regular arrangement of atoms which reflect light like tiny mirrors (see Figure 3.21). What happens when we shine light on crystals? We get smooth reflections, no diffraction or interference patterns. Now that we know something about diffraction and interference, we know that the *lack of* interference from atoms in a crystal implies something about the relationship between the distance between atoms and the wavelength of the light. Stop and think a minute to see if you can figure out this relationship.

If the distance (d) between atoms is just a little greater than the wavelength of visible light (λ), we would see diffraction and interference patterns. Since we don't see these patterns, the distance between atoms must either be much greater than a visible light wavelength or else less than a visible light wavelength. If atoms were considerably farther apart than a visible light wavelength, we'd be able to see the atoms under a strong microscope. Since we can't see atoms, even under the strongest microscopes, this implies that the distance between atoms must be less than the wavelength of visible light. This indicates that the reason we get straight reflections from crystals is that the atoms are closer together than the wavelength of visible light. Although this doesn't tell us how far apart the atoms *are*, it tells us how far apart they *are not*. They are not as far

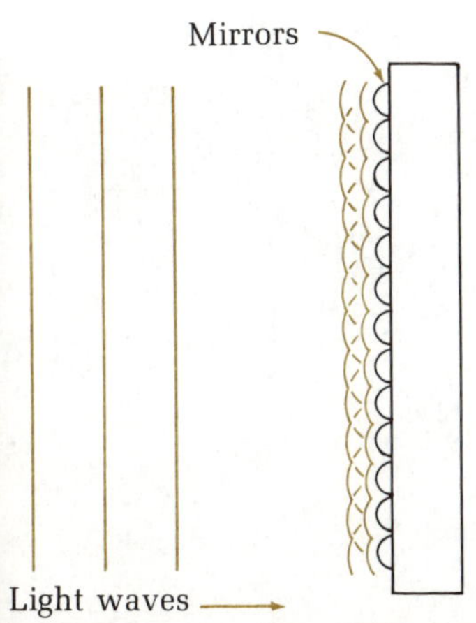

Figure 3.21 *Regular arrangement of tiny mirrors reflecting light*

apart as a wavelength of visible light. How can we go about using these diffraction and interference phenomena to explore crystals to find out how far apart the atoms actually are? Well, it would seem that what we need to look for is some shorter wavelength waves. While this looks all very reasonable now, it wasn't nearly as reasonable in 1912 when it was first done by some clever scientists.

3.3

The Von Laue Synthesis

It all came together at the University of Munich in 1912—the breakthrough that not only permitted an estimate of the distance between atoms, but also gave the scientific community an extremely accurate instrument to measure the distances involved. The breakthrough also incidentally proved the repetitive structure of crystal building blocks and the wave nature of x-rays. Sounds like a pretty big job, doesn't it? It was, but to stress the humanity of those involved, let's look at a sociodrama form of presentation of these events. (You might find it interesting to either write a script for this, or do it the "method" way by acting out the roles and speaking whatever seems appropriate.)

Von Laue Gets It Together—A Sociodrama

Scene #1: In 1895 in a small laboratory in Germany, Professor Wilhelm Roentgen (pronounced Rent'gen), after many years of tedious work with various tubes and apparatus, discovers a ray which seems to go right through things. He isn't sure what it is; so he calls it an x-ray.

Scene #2: In 1912 in a laboratory at the University of Munich, the full professor Roentgen (now about 65) explains his beloved x-rays to assistant professor Max Von Laue. Roentgen thinks they are electromagnetic waves like light, but cannot prove it.

Scene #3: In 1912 on a street in Munich, Max Von Laue (pronounced wow'uh) is walking along with a young Ph.D. candidate, Paul Ewald (pronounced ay'vald). Von Laue has been doing the usual teaching and research duties of a young professor struggling to become tenured, and is also working on an encyclopedia article about diffraction and interference. (Publish or perish, you know.) Ewald is writing his dissertation on the theory of the interaction of light with a regular arrangement of oscillators. Ewald's dissertation is entirely theoretical, but he suggests to Von Laue that his work might apply to light interacting with crystals. As Ewald discusses his thesis and makes this suggestion to Von Laue, Von Laue is clearly preoccupied with all his own problems and barely listens politely. Ewald finally senses Von Laue's preoccupation and gives up, leaving Von Laue to walk alone. As Von Laue continues by himself, the thought suddenly occurs to him that, although Ewald's suggestion about scattering light from crystals was ridiculous, because the wavelength of light was too large, *suppose Roentgen's x-rays were wavelike and had a wavelength about the same size as the distance between atoms in a crystal. The x-rays should diffract and cause interference patterns if they were shined on crystals.* He resolves to try some experiments along this line.

Scene #4: In the laboratory at the University of Munich, Von Laue is talking to two graduate students who are working as laboratory assistants, Friedrich and Knipping. He directs them to set up an experiment using Roentgen's x-ray tube and a crystal. They mutter a bit about the vagueness of the instructions and mess up the first try. Von Laue calls them dumkopfs, and they try again. This time their results show an interference pattern, and Von Laue is tremendously pleased.

Since crystals *do* cause interference patterns when x-rays are shined on them, this implies that *the distance between atoms in a crystal is slightly larger than the wavelength of an x-ray.* The experimental apparatus used to shine x-rays on crystals has been modified and improved to measure the distances between atoms in the building blocks of all known crystals. These techniques have been collected under the name *x-ray crystallography.*

Von Laue made an enormous contribution to our understanding of solid matter. With just one experiment, he

1. demonstrated the wave nature of x-rays, because x-rays

formed interference patterns, and only waves can interfere;

2. demonstrated the regularity of the arrangement of atoms, because only a regular arrangement could cause interference;
3. showed that the distance between atoms in a crystal is a little more than the wavelength of an x-ray;
4. supplied science with an experimental procedure for making very precise measurements on an atomic scale.

It's not difficult to see why Von Laue was awarded the Nobel Prize for science in 1914.

We can now answer the central question of this chapter: How far apart are the atoms arranged in the fundamental building block structure of a crystal? The atoms themselves are about 10^{-10} m in diameter, and x-ray crystallography tells us that their centers are about an x-ray wavelength, about 10^{-10} m, apart. Thus, the atoms in a crystal are arranged such that their electron clouds are nearly touching. Our picture of a cubic crystal is shown in Figure 3.22.

You may be asking yourself just what holds these atoms in these positions. Good question; let's take that up in the next chapter.

Summary

We began analyzing solid matter by melting, crystallizing, growing, and cleaving one form of solid—crystals. On the basis of our observations, we theorized that crystals must be built from fundamental building blocks of various geometric shapes. We then encountered some difficulty in determining the size relationship between atoms and these

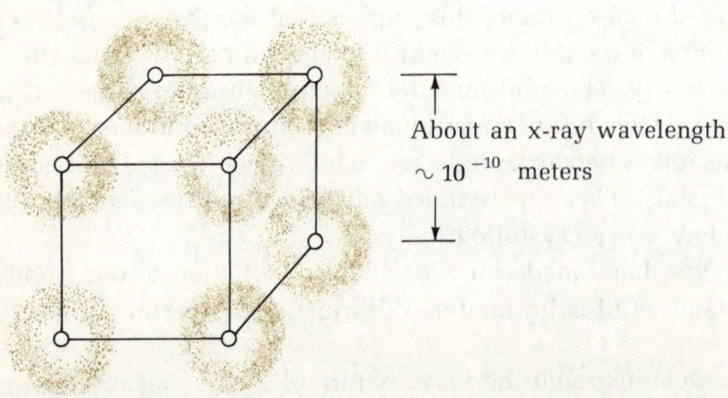

Figure 3.22 *Arrangement of atoms in a cubic crystal*

fundamental building blocks and in explaining why atoms don't cause wavy reflections from crystal surfaces. To resolve these difficulties, we began looking at light. The two main theories of light are the particle theory and the wave theory. The particle theory says that light consists of a stream of tiny particles; while the wave theory says that light consists of a wave similar to the water wave caused by dropping a stone into a pond. Created by the vibration of electric charges and depending on the charge's vibration frequency, a whole spectrum of electromagnetic waves may be generated. All electromagnetic waves travel at the same speed, c. An experiment on diffraction showed that light acts like waves when the size of the hole through which the light is shined begins to approach the wavelength of the light. Next, we looked at the phenomenon of interference, in which more than one wave occupies the same place at the same time. The resulting wave is found by simply adding the displacements of the interfering waves, according to the principle of superposition. When light goes through two slits which are close together, it diffracts at the slits and interferes beyond the slits, causing a pattern of alternating bright and dark lines. Doing some geometry revealed a quantitative relationship which was confirmed by an experiment. The quantitative relationship stated that interference patterns would be obtained only when the distance between slits was larger than, but not too much larger than, the wavelength of the light being used. Next, the two-slit apparatus was generalized to a series of slits; then the series of slits was replaced by a series of tiny mirrors, all of which would still yield interference patterns. When the mirrors were replaced by atoms, the lack of an interference pattern implied that the atoms must be closer together than one visible light wavelength. In 1912 this atomic spacing matter was settled by Von Laue when he generated interference patterns by shining x-rays—short wavelength, electromagnetic radiation—on crystals. Thus, we find that atoms are about an x-ray wavelength apart, which makes their electron clouds nearly touching as they are arranged in the fundamental building blocks of a crystal.

Questions

This may be used like a programmed text by referring to the indicated questions while reading the text.

3.1

1. What are the chief characteristics of crystals you observed in the first three experiments?
2. What leads you to believe that crystals are made of fundamental building blocks?
3. How can the fundamental building block idea be made to include atoms, which was the whole point of our study in Chapter 2?
4. What are the different possibilities for the relationship between atoms and building blocks?
5. How do smooth reflections from crystal faces enter our consideration of the relationship between atoms and building blocks?

3.2

6. What are the two competing theories of light?
7. What is the relationship between the motion of the medium and the motion of a transverse wave?
8. Identify the various parts of a wave.
9. What is the relationship between the period and the frequency of a wave?
10. Explain how the equation $c = f\lambda$ was arrived at.
11. Use the equation $c = 3 \times 10^8$ m/sec $= f\lambda$ to determine the wavelength of an infrared ray whose frequency is 6×10^{12} cycles/sec. (5×10^{-5} meters)
12. Use the equation $c = 3 \times 10^8$ m/sec $= f\lambda$ to determine the frequency of an x-ray whose wavelength is 1.5×10^{-10} meters. (2×10^{18} cycles/sec)
13. Explain the phenomenon known as diffraction.
14. Explain the connection between apparatus size and wavelength for diffraction.
15. Explain the phenomenon called interference.
16. Describe how waves come together according to the principle of superposition.
17. Explain the reason why you don't see the variation from light to dark as crests and nodes of light waves go by.
18. How does the wave theory prediction for the result of the double-slit experiment differ from the particle theory prediction?
19. Why can a bright spot be formed only from two crests (or two troughs) interfering, but not one crest and one trough?
20. Why must the light from one slit travel *exactly* one

wavelength farther than the light from the other slit in order to make a bright spot?

21. Explain each of the terms in the equation $\lambda/d = x/L$.
22. How can you tell from the equation that the pattern "spreads out" as d becomes closer to λ?
23. What would the interference pattern look like if there were a whole series of slits, rather than just two?
24. What would the interference pattern look like if there were a whole series of mirrors rather than slits?
25. If the mirrors were really atoms, for what relationship between d and λ could we expect to see interference patterns?
26. The fact that crystals do *not* yield interference patterns when visible light is shone on them tells us *what* about the relationship between the distance between atoms and the wavelength of visible light?

3.3

27. Describe briefly the role of each of the participants in the Von Laue synthesis.
28. What were the four main conclusions of the Von Laue synthesis?
29. How far apart are the atoms in a crystal?

Quotes to Ponder

1. "The crystal face of a mineral reflects its internal arrangement of atoms, much as a human face reflects the shape of the bones beneath it. A mineral the size of a pinhead might contain some eighty trillion atoms, arranged in a precise, three dimensional latticework." (J.A. Mandarino, "Fragments of Beauty from the Miniature World of Microcrystals", *Saturday Review,* February 19, 1972.)

CHAPTER FOUR

WHAT HOLDS THE ATOMS IN PLACE IN SOLID MATTER?

Now that we have determined that crystals consist of some orderly arrangement of atoms, and these atoms are not exactly touching each other but their centers are separated by a distance of about one x-ray wavelength, about 10^{-10} meters, we must ask, Why do these atoms want to hang around together? *Force,* you say? What exactly are forces, and how can they possibly work between things which are not "touching?" This idea of noncontact or action-at-a-distance forces had some of the old philosophers, like Aristotle, really perplexed. But we won't let this get in our way. First, we'll have a look at what Newton did with forces in general; then, we'll put some requirements on *the* force that holds solid matter together based on ordinary observations. We'll look at several possible candidates for *the* force; then we'll pick the winner by evaluating the candidate forces against the requirements.

4.1

Inertia, Force, and All That Newtonian Stuff

As you might surmise from Figure 4.1, the term *inertia* refers generally to a *resistance to change.* (Taking a course in physical science sometimes requires overcoming inertia.

Figure 4.1 *Inertia demonstration by L. Ump*

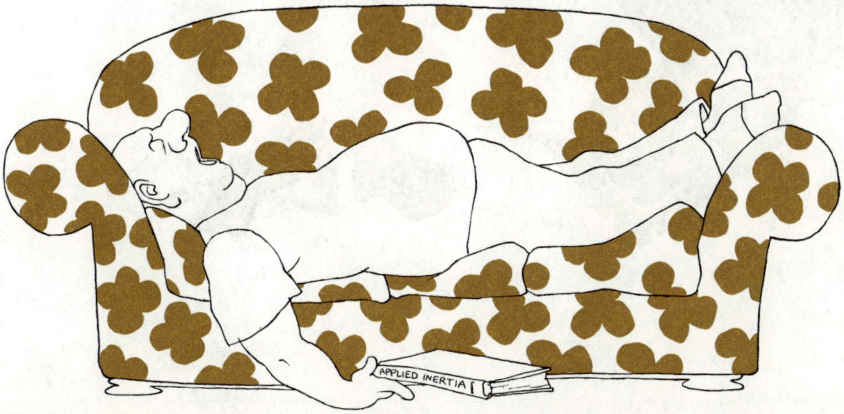

Congratulations.) Clearly, this has broad applicability in the intellectual realm, in which there is often a reluctance to accept new ideas, particularly in dealing with other people. In physical science, the term inertia refers to a resistance to change in the motion of bodies.

Newton's first law, the law of inertia, may be stated: *A body in motion in a straight line tends to stay in motion in a straight line; a body at rest tends to stay at rest until the body is acted upon by an outside force.*

To really understand this, we'll first have to think about motion and see some examples of Newton's first law.

Ever try to get some salt out of a balky salt shaker (Figure 4.2)?

Figure 4.2 *Salt coming out of a salt shaker*

The whole thing—salt, shaker, and all—is moving along at a constant velocity. Suddenly you stop the shaker, but the salt obeys Newton's first law and keeps right on going at its constant velocity; so it goes out the holes and onto your food.

Another example of Newton's first law in action is shown in the automobile-truck collision in Figure 4.3.

The truck and its occupant, B. Ump, are zipping along until they encounter a motionless car and its occupant. The truck's occupant tries to obey Newton's first law and to continue in motion. He winds up jamming his kneecap against the radio knobs and his head against the windshield.

Figure 4.3 *Auto collision involving B. Ump*

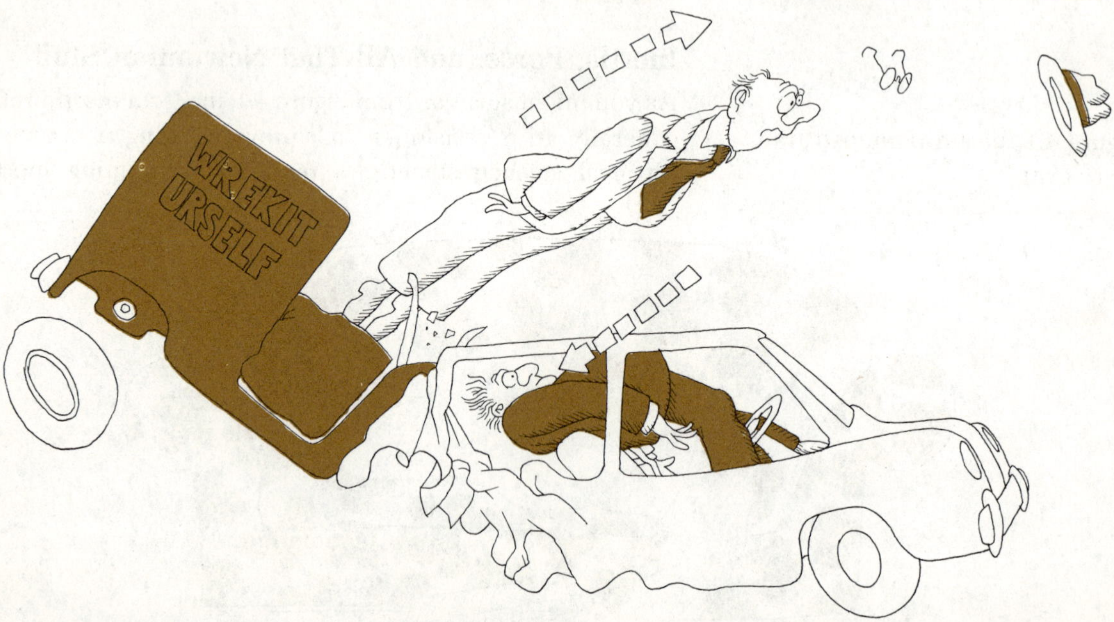

(Seatbelts and shoulder harnesses are provided so that the car and its occupants tend to move as a unit, rather than separately.) The car's occupant also obeys Newton's first law by trying to continue his state of nonmotion. The car, however, does move, so that his body gets squashed back into the seat, and his head is thrown back. (Head restraints are provided to convince your head to move along with the rest of the car.)

SUSPICIOUS STUDENT: We've gone several pages with no math. I think you must have some equation for us.

PLEASED PROFESSOR: I'm glad you've noticed by now that science must become quantitative so that measurements can be made to verify theories. Not to disappoint you, we'll introduce the quantity called *velocity* to make Newton's first law quantitative.

$$\text{Velocity} = \frac{\text{Distance traveled in a particular direction}}{\text{Time it takes to travel that distance}},$$

or, even better,

$$\text{Velocity} = \frac{\text{Distance}}{\text{Time}},$$

or, simply,

$$v = \frac{d}{t}.$$

In the MKS (meter, kilogram, second) system, the units of velocity are meters per second. Notice that, just as in several other equations we've seen, given any two of the quantities, it's possible to determine the third.

SUSPICIOUS STUDENT: Wait a minute! The signs I see on the highway say speed limit, not velocity limit. What's the difference between velocity and speed?

PROFESSOR: Velocity is a more precise concept because it involves motion *in a given direction,* while speed has no special direction associated with it. This distinction becomes important in curved motion. For example, 30 m/sec north is a velocity, but 30 m/sec with no direction specified is a speed.

To get back to velocity, let's do a couple of sample problems to make sure we understand.

A centimeter-worm can travel 20 cm in one minute. What is its speed, assuming it is uniform?

$$v = \frac{d}{t} = \frac{20 \text{ cm}}{60 \text{ sec}} = \frac{2 \times 10^{-1} \text{ m}}{6 \times 10^{1} \text{ sec}} = 3.33 \times 10^{-3} \text{ m/sec}.$$

A jetliner flies at 250 m/sec. How far does it go in 2 hours?

$$v = \frac{d}{t} \Rightarrow d = vt$$

$$= 250 \text{ m/sec} \times 2 \text{ hrs} \times 60 \frac{\text{min}}{\text{hrs}} \times 60 \frac{\text{sec}}{\text{min}},$$

$$= 5 \times 10^2 \times 36 \times 10^2 = 1.8 \times 10^6 \text{ m}.$$

Your auto zips along the highway at about 30 m/sec. How long would it take to go 60 kilometers?

$$v = \frac{d}{t} \Rightarrow t = \frac{d}{v} = \frac{60 \times 10^3 \text{ m}}{30 \text{ m/sec}} = 2 \times 10^3 \text{ sec} = 33 \text{ min}.$$

The quantity which resists change, according to Newton's first law, is mass times velocity, called *momentum*. For a body whose mass doesn't change, this implies that the constant quantity is the velocity.

SHARP STUDENT: What you're trying to tell me is that for moving bodies which neither lose nor gain mass, the quantity which resists change is the velocity. In other words, Newton's first law says that bodies tend to continue to move as fast as they were and in the same direction until a force acts to change this.

PLEASED PROFESSOR: I do believe you've got it. Now, let's go on and investigate what happens when velocity changes. We'll start by defining a new term, *acceleration*.

$$\text{Acceleration} = \frac{\text{Amount of change in velocity}}{\text{Amount of time over which velocity changed}},$$

or, better still,

$$\text{Acceleration} = \frac{\text{Change in velocity}}{\text{Change in time}},$$

or, simply,

$$a = \frac{\Delta v}{\Delta t},$$

where Δ (Greek letter delta) stands for change in whatever follows.

CLEVER STUDENT: I knew there would be an equation in there somewhere.

PROFESSOR: You bet. The units of acceleration are liable to confuse you; so watch closely:

$$a = \frac{\Delta v}{\Delta t} = \frac{\text{m/sec}}{\text{sec}} = \text{m/sec}^2.$$

SUSPICIOUS STUDENT: A sec^2? Is that longer or shorter than a second?

PROFESSOR: A sec^2 is not a unit of time, so it cannot be compared to a second. It is a product of two time units, one coming from the change in velocity (m/sec) and the other coming from the time interval over which the velocity changes (sec).

Perhaps an example or two will help straighten this out.

If your auto goes from 0 to 25 m/sec in 10 sec, what is its acceleration?

$$a = \frac{\Delta v}{\Delta t} = \frac{25\ \text{m/sec}}{10\ \text{sec}} = 2.5\ \text{m/sec}^2.$$

A jet airplane is capable of accelerating at a rate of 8 m/sec^2. How long would it take to go from 100 m/sec to 500 m/sec?

$$a = \frac{\Delta v}{\Delta t} \Rightarrow \Delta t = \frac{\Delta v}{a} = \frac{(500 - 100)(\text{m/sec})}{8\ \text{m/sec}^2} = 50\ \text{sec}.$$

IMPATIENT STUDENT: I know about all that; I go to the drags every once in a while. When are we going to get around to talking about force?

PROFESSOR: Right now. We have seen that Newton's first law says that the velocity of a body is constant unless some outside force is applied. We have seen that an acceleration is a change in velocity divided by a change in time. It seems fairly logical that an acceleration must be caused by a force. Now, what exactly is a force? Your intuition tells you that *a force is some kind of a push or pull,* and your intuition is right!

While we're exercising your intuition, how would you expect the force and the acceleration to be related? That is, if you apply more force to a body, should you get more acceleration, or less?

STUDENT: It seems to me that if I were pushing a car to get it started, the more I wanted to change the velocity, the more force it would take; so I'd say the more force, the more acceleration.

PROFESSOR: Correct. Using your car-pushing analogy, you'd certainly rather push a light car than a heavy one, so you'd probably say that force would have to increase as mass increases, also.

Newton took care of all this very neatly and compactly in *Newton's second law:*

$$\text{Force} = (\text{mass})(\text{acceleration}),$$

or,

$$\boldsymbol{F = ma.}$$

You see, force must increase as both mass and acceleration increase, just as your intuition expected.

The units of force may be found simply by multiplying the units for mass and acceleration:

$$F = ma = \text{kg}\,\frac{\text{m}}{\text{sec}^2}.$$

The unit of force, the kg m/sec^2, is called a *newton* to honor Sir Isaac Newton. (We scientists take care of our own.)

$$1\,\frac{\text{kg m}}{\text{sec}^2} = 1 \text{ newton} = 1 \text{ nt}.$$

STUDENT: Could we have some sample problems to work to help us understand this concept?

PROFESSOR: Delighted to oblige. It usually helps to draw a little sketch of the object, putting in whatever forces are acting on it. Then we can find the net force and go from there.

SUSPICIOUS STUDENT: Hold it a minute—I mean, excuse me, but what are you saying about a net?

PROFESSOR: My fault. If there are two or more forces in a given direction, you have to add them up to get the net force (see Figure 4.4).

STUDENT: Suppose there was as much force one way as the other?

PROFESSOR: Well, there would be zero net force and so . . .

STUDENT: . . . and so there would be zero acceleration, which means no change in velocity. This brings us right back to Newton's first law.

PLEASED PROFESSOR: Beautiful! In a sense, Newton's first law is just a special case of Newton's second law. No net force → no acceleration → no change in velocity. Bodies in motion tend to stay in motion; bodies at rest tend to stay at rest.

Let's see, we were going to look at some $F = ma$ examples.

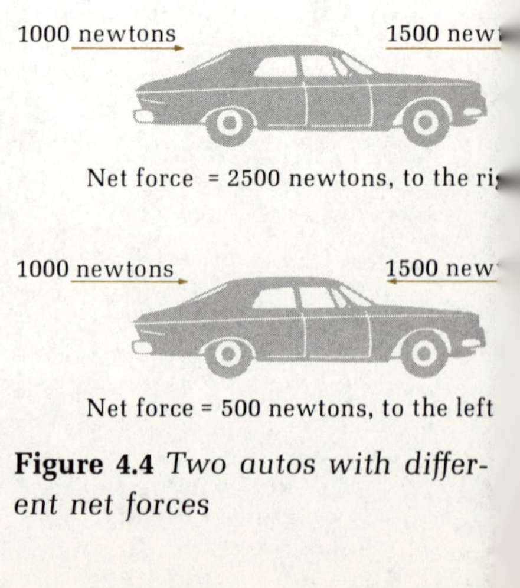

Figure 4.4 *Two autos with different net forces*

Figure 4.5 *Iceberg accelerated by ship*

If a force of 4×10^8 newtons is applied to a floating iceberg by an unwary ship, how fast will the 2×10^{10} kg iceberg accelerate away, neglecting water resistance (Figure 4.5)?

$$F = ma \Rightarrow a = \frac{F}{m} = \frac{4 \times 10^8 \text{ nt}}{2 \times 10^{10} \text{ kg}} = 2 \times 10^{-2} \text{ m/sec}^2.$$

If a 4 kg baseball is observed to accelerate at a rate of 30 m/sec^2, what net force is acting on this ball (Figure 4.6)?

$$F = ma = (4 \text{ kg})(30 \text{ m/sec}^2) = 120 \text{ nt}.$$

A prime example of one particular force which acts on us all the time is the gravitational force. Newton's second law takes the special form

$$w = mg,$$

where w = weight (newtons),
m = mass (kilograms),
g = acceleration due to gravity $\approx$ 10 m/sec^2 at the earth's surface.

Thus, a person whose mass is 50 kg would have a weight of 500 newtons at the earth's surface.

STUDENT: But I thought mass and weight were the same thing.

PROFESSOR: Not at all! The *mass* is a fundamental quantity measuring the amount of *matter* or "stuff" a body possesses, and it has units of kilograms. The *weight*, on the

Figure 4.6 *Baseball accelerated by a bat*

other hand, is a measure of the gravitational *force* being exerted on the object, and it has units of newtons.

STUDENT: Oh, I get it. Astronauts are *weightless* in space, and they weigh less on the moon than on earth, but these terms refer only to the *force* on them, not the amount of matter their bodies contain.

PROFESSOR: Precisely true. The mass is thus a more fundamental quantity because it remains the same wherever an object is, while the weight may vary, depending on how far from the planet the object is.

By the way, speaking about weight, mass, and the acceleration due to gravity, we're about to do an experiment that should enable us to measure g, shoot down Aristotle's theory about heavy objects accelerating faster than light ones, and just generally firm up the whole Newtonian business for you.

EXPERIMENT 4.1
MEASUREMENT OF *g*, THE ACCELERATION DUE TO GRAVITY

What goes up must come down and all that jazz. Anything that falls is being acted upon by the force of gravity, and

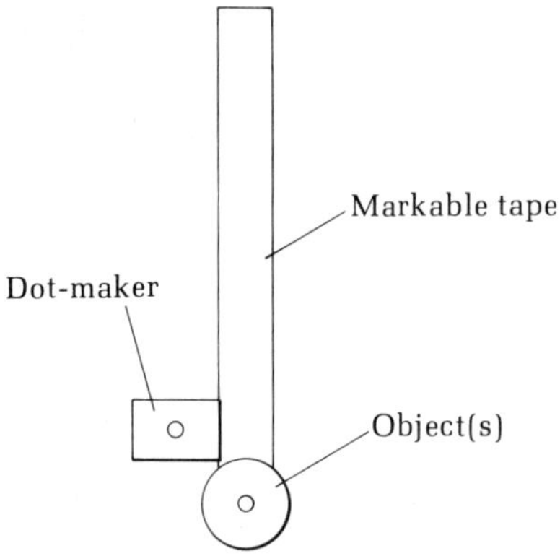

Figure 4.7 *Apparatus for g-measurement*

so it accelerates. The acceleration which any falling body has is said to be independent of the mass of the body, and this acceleration should have a value of about 10 m/sec^2 near the surface of the earth. Of course, we don't have to accept anyone's word for this; we can prove it according to the ultimate authority in physical science—experiment.

Since the Leaning Tower of Pisa is not available for us to go drop some cannonballs, as Galileo did, we'll have to try dropping something a little simpler in the laboratory. We'll measure some distances and times, calculate the velocity, calculate the acceleration, and compare it for two objects of different mass. To get the velocities, we'll need some way of measuring distances and times (see Figure 4.7).

Procedure

1. *Attach the markable tape to the objects to be dropped, and feed the tape through the dot-maker.*
2. *Turn on the dot-maker and let go of the markable tape, allowing the object(s) to fall, dragging the markable tape through the dot-maker.*
3. *The dot-maker will make a bunch of dots on the tape which may look something like that illustrated in Figure 4.8.*
4. *You can measure the distance between dots with a meter stick; and if you know the time between dots, you can calculate the velocity easily enough:*

$$\text{Velocity} = \frac{\text{Distance between dots}}{\text{Time interval between dots}}.$$

For example, if the dot-maker is going at 30 dots/second, there must be a 1/30 second time interval between one dot and the next; so, over the first time interval, the velocity is the 2 cm measured distance between dots divided by the 1/30 second time interval: $v = 2\text{ cm}/(1/30\text{ sec}) = 60\text{ cm/sec}$. See Table 4.1. (Round off velocities to the nearest 1 cm/sec or you'll be carrying along too many numbers.)
5. *Now that you have the velocity history calculated, a good way to understand what you've got is to draw a graph of it (see Figure 4.9).*

This graph is a little weird, however. Most graphs have points rather than lines like we have. Where should the velocity point go? It can't go to the beginning of the interval because it didn't really start out going this fast; nor can it go to the end of the interval, because by then it is going even faster. So it must go in the middle, because it is really

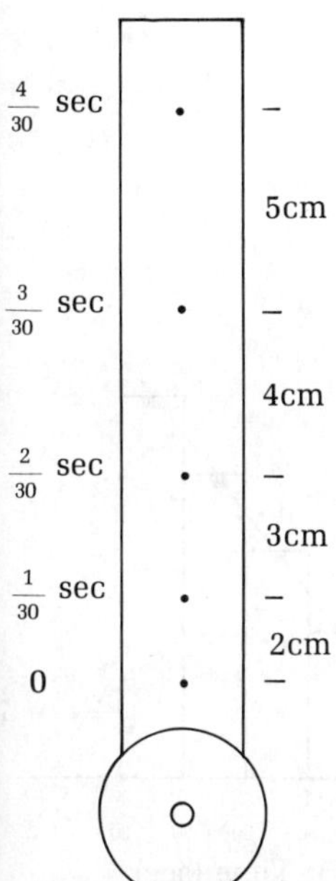

Figure 4.8 *Sample tape for g-measurement*

Table 4.1

Time Interval From	To	Time between Dots	Distance	Velocity
0	→ $\frac{1}{30}$ sec	$\frac{1}{30}$ sec	2 cm	60 cm/sec
$\frac{1}{30}$ sec	→ $\frac{2}{30}$ sec	$\frac{1}{30}$ sec	3 cm	90 cm/sec
$\frac{2}{30}$ sec	→ $\frac{3}{30}$ sec	$\frac{1}{30}$ sec	4 cm	120 cm/sec
$\frac{3}{30}$ sec	→ $\frac{4}{30}$ sec	$\frac{1}{30}$ sec	5 cm	150 cm/sec

an average velocity. In that case, put the dots at the middle of each time interval for the average velocity and draw a single, continuous, straight line that comes as close as possible to going through all the average velocity points.

6. *The acceleration can now be found by simply recalling the equation:*

$$\text{Acceleration} = \frac{\text{Change in velocity}}{\text{Change in time}} = \frac{\Delta v}{\Delta t}.$$

We have a graph showing velocity vs. time; so we can take any two points on the straight line and calculate the acceleration. For example, taking the first and fourth points on our line,

$$\text{Acceleration} = g = \frac{150\ \text{cm/sec} - 60\ \text{cm/sec}}{\frac{3\frac{1}{2}}{30}\ \text{sec} - \frac{\frac{1}{2}}{30}\ \text{sec}},$$

$$= \frac{90\ \text{cm/sec}}{\frac{3}{30}\ \text{sec}} = 900\ \text{cm/sec}^2 = 9\ \text{m/sec}^2.$$

7. *Compare this value with the expected result, 10 m/sec² to see what the percent difference is.*

$$\text{Difference} = \text{Expected value} - \text{Experimental value},$$

$$10\ \text{m/sec}^2 - 9\ \text{m/sec}^2 = 1\ \text{m/sec}^2,$$

$$\text{Percent difference} = \frac{\text{Difference}}{\text{Expected value}} \times 100\%$$

$$= \frac{1\ \text{m/sec}^2}{10\ \text{m/sec}^2} \times 100\% = 10\%.$$

(Note: The experimental value is lower. What is there in the experiment setup to cause you to expect this result?)

8. *Now repeat the experiment with an object of different mass and calculate its acceleration. Compare the two*

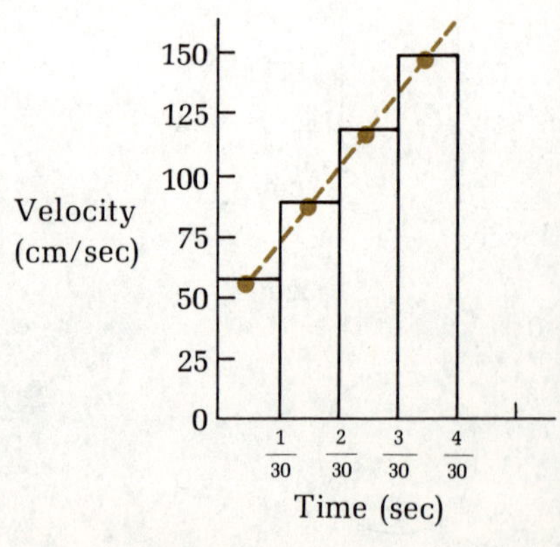

Figure 4.9 *Example of velocity history*

Figure 4.10 *Arrangement of atoms in a crystal of NaCl*

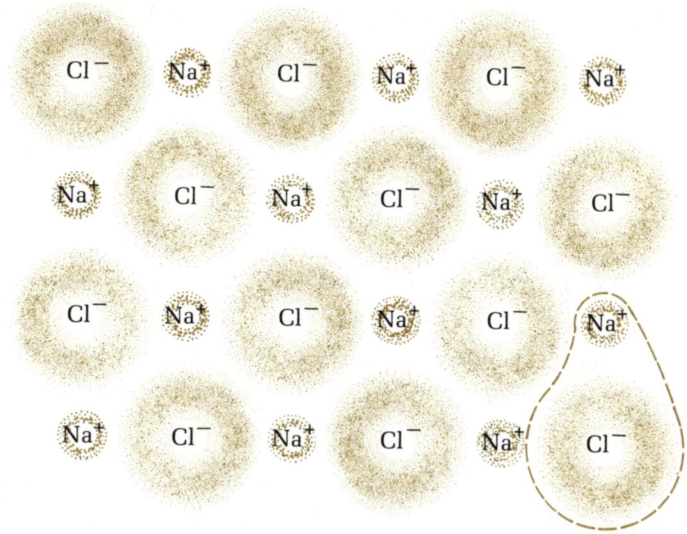

values of g. Either they should agree with each other fairly well, or you should have some good reason why they don't.

4.2

Some Intuitive Requirements on the Force

Now that we have come to grips with force, we should be able to focus in on the atoms in a crystal of solid matter and place some intuitive requirements on the force which holds the atoms together in a regular array (see Figure 4.10).

We can make some fairly commonplace observations which have implications on *the* force which keeps these atoms in their appropriate spots.

1. *It is not easily possible to either pull the crystal apart, or to squash it together.* That's the whole point about solids—they maintain their shape without any help or support from any container.

   Looking at our two atoms (Figure 4.11), this observation implies that there must be (a) an *attractive force* keeping them from being pulled apart and (b) a *repulsive force* keeping them from being squashed together. This implies that *the* force must be capable of being both *attractive* and *repulsive*.

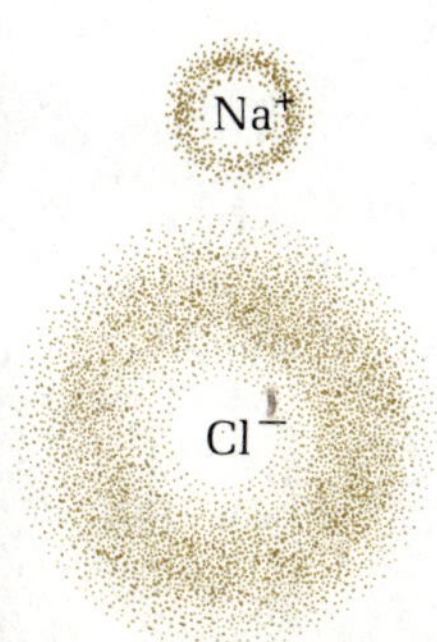

Figure 4.11 *The bottom two atoms in the crystal of NaCl*

2. *The bottom atoms don't fall off under the influence of gravity.* Looking at just the bottom atom alone, there are two forces acting on it: its *weight* pulling it down, and the force exerted on it by the atom above (Figure 4.12).

   In order for the bottom atom to *not* fall off, *the* force, *F*, acting upward must be *greater than or equal to the weight* which acts downward.
3. *There are very many different kinds of solid matter.* It would be simplest if we could find just *one* force which would work for all the different kinds of solid matter. This requires that our candidate force be *universal,* that is, it must work on all kinds of solid matter.

To summarize these observations:

Requirement	Reason
The force must be:	
1. Both attractive and repulsive	1. It is not easily possible to either pull the crystal apart, or squash it together.
2. Greater than the weight of an atom	2. The bottom atoms don't fall off under the influence of gravity.
3. Universal	3. One force should work for all matter for simplicity.

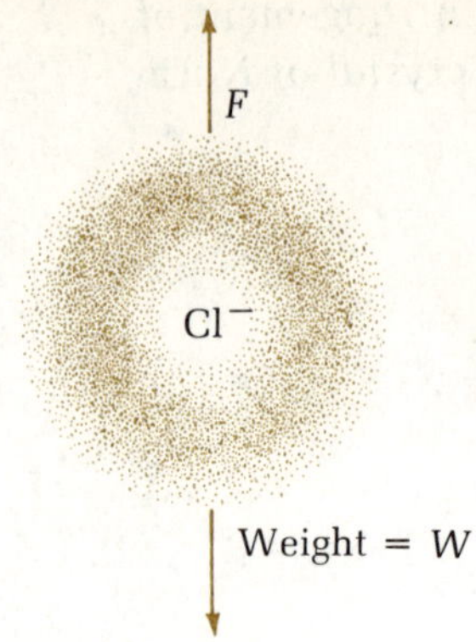

Figure 4.12 *Forces on the bottom atom*

4.3

Possible Candidates for "the" Force

Now that we know something about forces in general, and have made some observations which place requirements on *the* force which holds solid matter together, let's look at some different kinds of forces to see how they work and how well they meet these requirements. (We'll see these forces again in Chapter 6, when we try to figure out how the universe works.)

Gravitational Force

Newton formulated the law of universal gravitation, according to the old story, while sitting under an apple tree. Per-

Figure 4.13 *Newton and the old apple*

haps Figure 4.13 illustrates how it really happened.

Newton postulated that any two bodies that have mass attract each other according to the following relation (see Figure 4.14):

$$F_G = G\frac{m_1m_2}{r^2},$$

where

$$G = \text{universal constant},$$

$$= 7 \times 10^{-11}\ \frac{\text{nt}\cdot\text{m}^2}{\text{kg}^2},$$

$$m_1, m_2 = \text{masses (kg)},$$

$$r = \text{distance between centers of masses (meters)}.$$

This is not too difficult to believe, because we would intuitively expect two bodies to exert more force on each other as their masses increase, and we would expect the force to decrease as the distance between the objects increases (out of sight, out of mind).

Let's work some examples to see how this force operates.

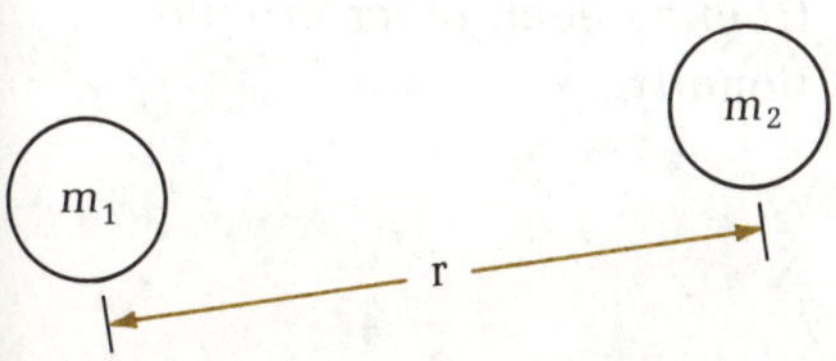

Figure 4.14 *Objects exerting a gravitational force on each other*

Example #1: What is the gravitational attractive force between two automobiles, each having 2000 kg mass, parked in a parking lot, with their centers 3 meters apart (Figure 4.15)?

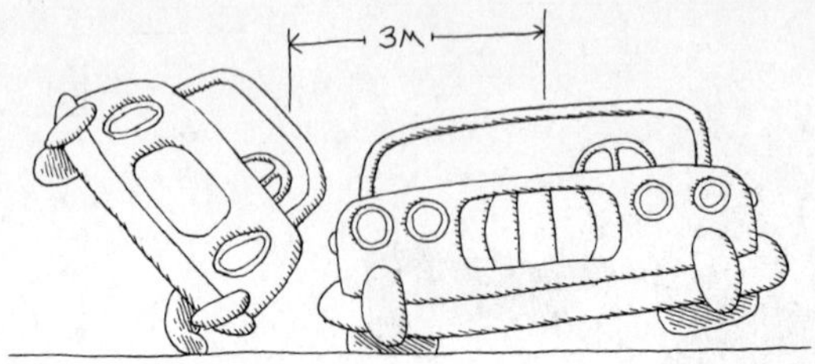

Figure 4.15 *Two autos attracting each other gravitationally*

$$F_G = G\frac{m_1 m_2}{r^2},$$

$$F_G = 7 \times 10^{-11}\frac{\text{nt}\cdot\text{m}^2}{\text{kg}^2}\frac{(2000\text{ kg})(2000\text{ kg})}{(3\text{ m})^2},$$

$$= \frac{7 \times 10^{-11}(2 \times 10^3)(2 \times 10^3)}{9}\text{ nt},$$

$$= \frac{28}{9} \times 10^{-5}\text{ nt} = 3.1 \times 10^{-5}\text{ nt}.$$

There is not much force—about equivalent to the weight of a fly—and the friction is too large to overcome, so the cars don't move toward each other.

Example #2: How much gravitational force is exerted on the earth by the sun, and on the sun by the earth (Figure 4.16)?

$$F_G = G\frac{m_1 m_2}{r^2},$$

$$= 7 \times 10^{-11}\frac{\text{nt}\cdot\text{m}^2}{\text{kg}^2}\frac{(2 \times 10^{30}\text{ kg})(6 \times 10^{24}\text{ kg})}{(1.5 \times 10^{11}\text{ m})^2}$$

$$= \frac{7(2)(6)}{2.25} \times 10^{-11+54-22}\text{ nt},$$

$$= 37 \times 10^{21}\text{ nt} = 3.7 \times 10^{22}\text{ nt}.$$

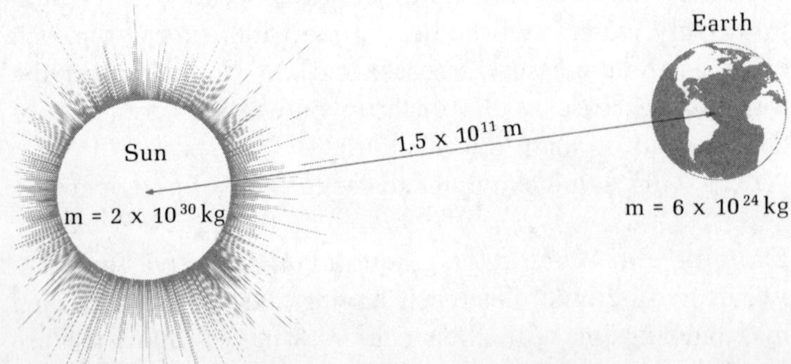

Figure 4.16 *Sun and Earth attracting each other gravitationally*

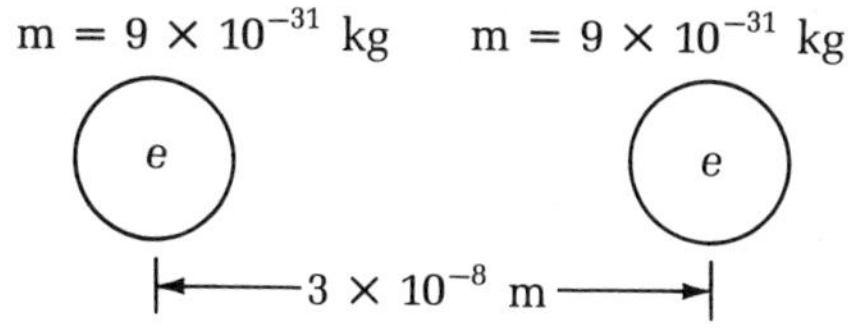

Figure 4.17 *Two electrons attracting each other gravitationally*

This is the force that keeps the earth whirling around the sun, rather than flying straight off as the law of inertia would demand. This force fulfills the same role as the string with which you twirl a stone or a yo-yo in a circular path.

Example #3: How much gravitational force is there between two electrons 3×10^{-8} meters away from each other (Figure 4.17)?

$$F_G = G\frac{m_1 m_2}{r^2},$$

$$= 7 \times 10^{-11}\,\frac{\text{nt}\cdot\text{m}^2}{\text{kg}^2}\,\frac{(9 \times 10^{-31}\,\text{kg})(9 \times 10^{-31}\,\text{kg})}{(3 \times 10^{-8}\,\text{m})^2},$$

$$= \frac{7(9)(9)}{9} \times 10^{-11-31-31+16}\,\text{nt},$$

$$= 63 \times 10^{-57}\,\text{nt} = 6.3 \times 10^{-56}\,\text{nt}.$$

Electric Force

Masses attracting each other are not difficult to visualize, but try to conjure up *charges* both attracting and repelling each other. (Careful, or you'll get your mind all bent out of shape.) Unfortunately, you can't see charges, but you can certainly see their effects. For example, if you were to rub plastic over an amber rod, then suspend the rod as shown in Figure 4.18, another amber rod rubbed with plastic would cause the suspended one to move away.

If you think that's strange, a glass rod rubbed with silk will attract the suspended amber rod.

For quite some time, electrical phenomena were regarded as mystical, but a French engineer, Charles Coulomb, put an end to the mystery with his hypotheses that: (1) Electricity comes in two varieties, which Ben Franklin called positive and negative. (The two kinds of charge could have been called almost anything; the point is that they are different.) (2) The force exerted by one charge on another is given by the equation

$$F_E = K\frac{Q_1 Q_2}{r^2},$$

where K = universal constant = $9 \times 10^9\,\text{nt}\cdot\text{m}^2/\text{coul}^2$,
Q_1, Q_2 = charges, measured in coulombs (*new fundamental unit*),
r = distance between charges.

(3) Like charges repel, while unlike charges attract.

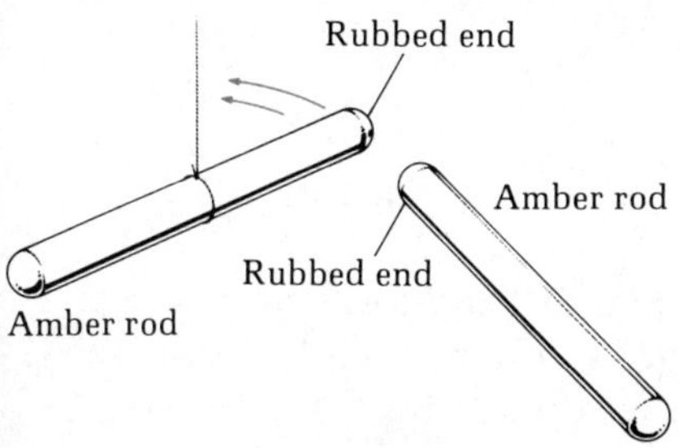

Figure 4.18 *Amber rods, rubbed with plastic, repelling*

So, the reason the amber rods rubbed with plastic repelled

each other was that they were both charged the same way (negatively), while the glass rod rubbed with silk was charged positively and therefore attracted.

Notice that the electric force introduced a new fundamental unit. Our old Meter Kilogram Second system must be expanded to MKSC (coulomb) to handle electricity. Thus, the coulomb is the fundamental unit of charge, just as the meter is the basic unit of length, the kilogram is the basic unit of mass, and the second is the basic unit of time.

Let's first work with the equation, then we'll do an experiment series to convince ourselves of the reality of charge.

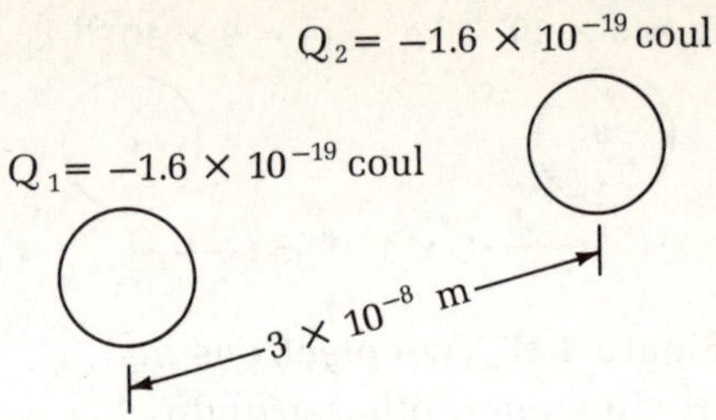

Figure 4.19 *Two electrons repelling each other electrically*

Example #1: The charge of the electron is -1.6×10^{-19} coulombs, as determined by an ingenious experiment done by Robert A. Millikan. Let's find the amount of electrical repulsion between 2 electrons that are 3×10^{-8} meters apart (see Figure 4.19).

$$F_E = K\frac{Q_1Q_2}{r^2},$$

$$= 9 \times 10^9 \frac{\text{nt} \cdot \text{m}^2}{(\text{coul})^2} \cdot \frac{(-1.6 \times 10^{-19}\text{ coul})(-1.6 \times 10^{-19}\text{ coul})}{(3 \times 10^{-8}\text{ m})^2},$$

$$= \frac{9(1.6)(1.6)}{9} \times 10^{9-19-19+16}\text{ nt},$$

$$= 2.56 \times 10^{-13}\text{ nt}.$$

Recall our example calculation where the gravitational force of attraction between two electrons yielded a force of 6.3×10^{-56} newtons. In Figure 4.20, let's look at one of the electrons subject to these two forces. It's pretty clear that the electric force vastly overpowers the gravitational force.

Example #2: In the gravitational-force examples, it was easy enough to get some large lumps of mass to exert quite

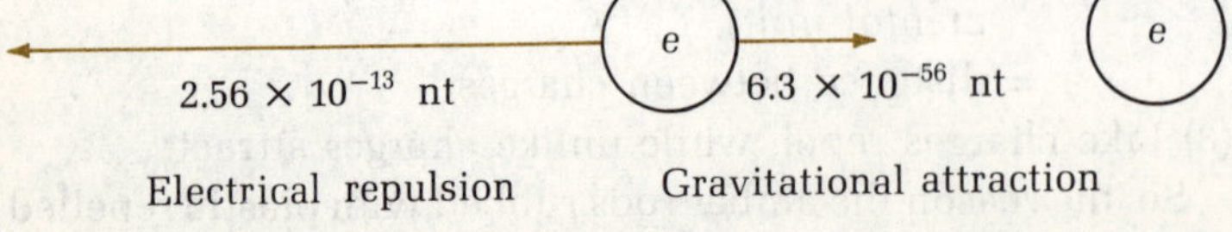

Figure 4.20 Forces on electron *(not drawn to scale)*

a bit of force on each other over fairly long distances. However, it's quite difficult to get large lumps of charge. Suppose we could get +1 coulomb and −1 coulomb within 100 meters of each other (a distance slightly longer than the length of a football field). What would be the electrical force of attraction?

$$F_E = K\frac{Q_1Q_2}{r^2} = 9 \times 10^9 \frac{\text{nt} \cdot \text{m}^2}{\text{coul}^2} \cdot \frac{(1)(1)\ \text{coul}^2}{(10^2)^2\ \text{m}^2},$$
$$= 9 \times 10^5 \text{ newtons.}$$

This is a force equivalent to the weight of about 50 automobiles!

It certainly should have become clear by now that the electric force is indeed quite strong. The gravitational and electric forces are quite similar in their formulation.

$$F_G = G\frac{m_1m_2}{r^2};$$
$$F_E = K\frac{Q_1Q_2}{r^2}.$$

They both consist of a constant, multiplied by the product of two fundamental quantities (mass or charge) and divided by the square of the distance between them. The two constants have grossly different numerical values, however.

$$G = 7 \times 10^{-11} \frac{\text{nt} \cdot \text{m}^2}{\text{kg}^2};$$
$$K = 9 \times 10^9 \frac{\text{nt} \cdot \text{m}^2}{\text{coul}^2}.$$

With the constant differing by a factor of more than 10^{20}, we can expect some great difference in the resulting forces.

To get electricity down to an even more understandable plane, let's do several small experiments which we'll place under one heading, called:

EXPERIMENT 4.2

GETTING TO KNOW ALL ABOUT ELECTRICITY

Static electricity experiments don't always work out precisely the way they are supposed to, mostly because of atmospheric conditions; so please have a lot of patience with these experiments.

A. Does It Work on a Lot of Different Things?

Put a variety of small solids into paper cups, e.g.: coffee,

Figure 4.21 *Plastic-rubbed amber rod attempting to pick up various small solids*

tea, sugar, salt, bits of paper, bits of aluminum foil, small pieces of plastic, breakfast cereal, pipe tobacco, pencil shavings, (Figure 4.21). Bring the amber rod rubbed with plastic above each one and see if it will pick anything up. Does this work for everything?

B. Induced Charge

1. *Don't rub the amber rod, but bring it near the pith ball (Figure 4.22) Does it have any effect? It shouldn't; but if it does, touch both the rod and the ball and try again. Neither should be charged, so there should be no effect.*

2. *Now rub the amber rod with the plastic and bring the rod near the pith ball, but <u>do not let them touch.</u> Now what happens? The pith ball should move toward the rod.*

3. *Now touch the rod to the ball (several times, perhaps) and notice what happens. The pith ball should move away from the rod.*

4. *To really boggle your mind, try to explain the entire sequence you just observed, using Coulomb's law,* $F_E = K(Q_1Q_2)/r^2$. *You should be in real trouble because the initial charge on the ball is zero, so the force should be zero also. But it obviously isn't zero because the rod and ball are attracted. Wrestle with this one a bit before you look at the explanation at the end of the next part of this experiment.*

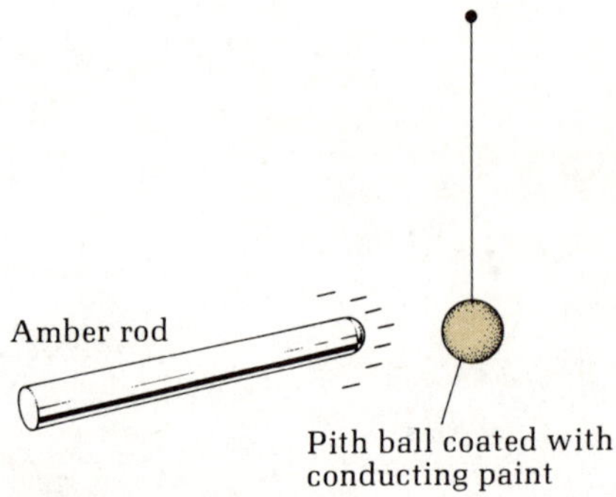

Figure 4.22 *Amber rod and pith ball setup*

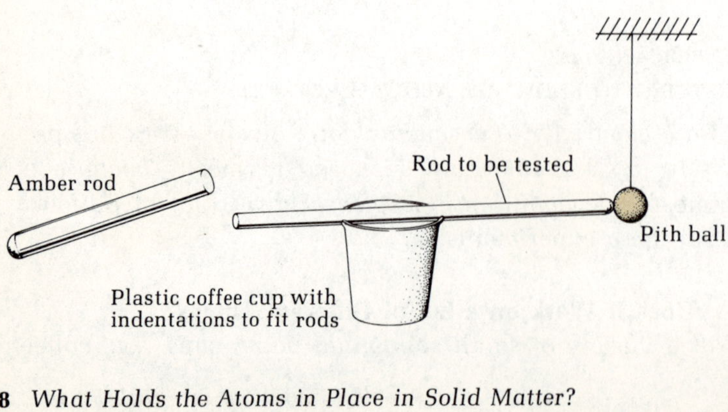

Figure 4.23 *Charge transfer experimental setup*

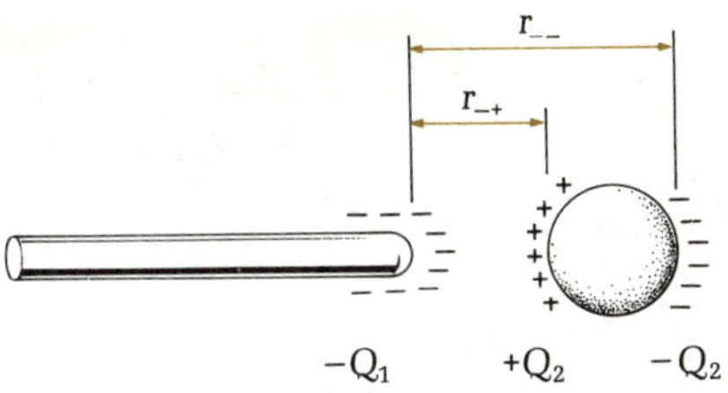

Figure 4.24 Induced charge diagram—uncharged ball

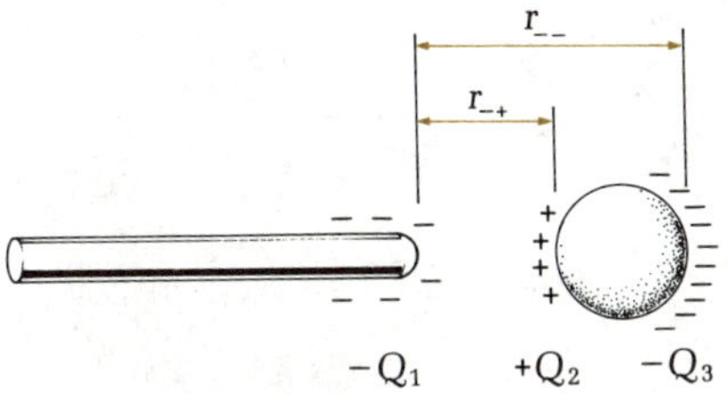

Figure 4.25 Induced charge diagram—charged ball

C. Transfer of Charge

Put various rods (glass, plastic, metal, chalk, a pencil sharpened at both ends, etc.) between the amber rod and the pith ball (Figure 4.23). Charge the amber rod, touch it to one end of the test rod, touch the pith ball to the other end. Note the action of the pith ball.

You should observe that some materials allow charge to flow through them rather easily, while others don't. Materials that allow charge to flow are called <u>*conductors*</u>*, while materials that don't allow charge to flow are called* <u>*non-conductors*</u>*. (Hint: To help explain part B, the pith ball in part B was painted with a conducting paint which allowed charges to move freely about on its surface.)*

Explanation of Part B

Although the pith ball is initially uncharged, it contains an equal number of positive and negative charges. The presence of the negatively charged amber rod repels the negative charges and attracts the positive. The conducting paint on the ball's surface allows the charges to move around and to rearrange themselves as illustrated in Figure 4.24.

Figure 4.26 Electromagnetic crane

The ball is still neutral overall, because it has just as many positive as negative charges. However, because the + and − are closer than the − and −, Coulomb's law becomes

$$F_{E\,\text{Attraction}} = K\frac{Q_1Q_2}{(r_{+-})^2};$$

$$F_{E\,\text{Repulsion}} = K\frac{Q_1Q_2}{(r_{--})^2}.$$

Because r_{+-} is less than r_{--}, the force of attraction is larger than the force of repulsion, so there is a net attraction initially.

Once the rod and ball touch, the ball becomes negatively charged. (Actually, electrons, which are negatively charged and extremely light, jump to the ball, making it negatively charged.) Now the ball and rod are both negative, and even though the charges rearrange themselves as before, the greater number of negative charges on the ball causes a net repulsion (Figure 4.25):

$$F_{\text{Attraction}} = K\frac{Q_1Q_2}{(r_{+-})^2};$$

$$F_{\text{Repulsion}} = K\frac{Q_1Q_3}{(r_{--})^2}.$$

Even though the distance r_{+-} is less than the distance r_{--}, Q_3 has become so much larger than Q_2 (because of charge transfer) that there is now a net repulsion.

Figure 4.27 *Unfortunate demonstration of the nuclear force*

Magnetic Force

Most people who have seen a compass are aware that there is such a thing as magnetic force; therefore, we had better look at magnetic force as a potential candidate for *the* force which holds solid matter together.

Some horseshoe magnets are available which can lift almost 100 times their own weight. Electromagnets have so much lifting power they are used in junkyards to rearrange the rusting automobile hulks to make room for more (see Figure 4.26 on the preceding page).

The mathematical laws which describe the magnetic force are really too complicated to go into here, but there are a few features we should be aware of:

1. The magnetic force from a small source varies with dis-

tance approximately the same way as both the gravitational and electric forces.

2. There are two poles, called *north* and *south*, with like poles repelling each other and unlike poles attracting.
3. Magnetic force works only on certain metals, notably iron and to some extent nickel and gadolinium.
4. All magnetic behavior can be linked to moving charges, leading to the idea that electricity and magnetism are tied together so intimately that they are jointly known as the electromagnetic field.

Nuclear Force

Another force whose strength is clearly unquestionable is the nuclear force. Unfortunately, the power of this force was demonstrated quite graphically in 1945 (see Figure 4.27). (Sorry, but we're not going to do any experiments in this section.)

The mathematical statement of the nuclear force is quite complex and not quite settled yet, but there are a few things that can be stated:

1. The nuclear force is rather short-range compared to gravitational, electric, and magnetic forces. To a certain degree of approximation,

$$F_{\text{Nuclear}} \propto \frac{1}{r^{15}}.$$

The r doesn't have to get very large for $1/r^{15}$ to get pretty small.

2. Protons and neutrons, but not electrons, are affected by the nuclear force.
3. The nuclear force is attractive only until extremely short distances are reached ($\sim 10^{-15}$ m), and then it becomes repulsive.

Other Forces

Most other forces that come to mind, like contact forces or frictional forces, are actually just different manifestations of *the* force holding solid matter together. The contact force is just the repulsive part of *the* force, refusing to allow the atoms to be squeezed together, while the frictional force is just the attractive part of *the* force, refusing to allow the atoms to be ripped off.

Actually, there is one other force recognized by scientists, called the *"weak" force*. It is stronger than the gravitational force but weaker than either the electromagnetic or nuclear forces. It is not at all well understood, but what little is known is quite alarming in the sense that a number of traditional theories have been shattered by the "weak" force. The situation is quite comparable to the ultraviolet catastrophe from Chapter 2. Physics is awaiting a Max Planck to clear away the confusion which presently exists in the world of fundamental particles and the "weak" force. (See Chapter 10.)

Since we don't know enough about it, we will try forces we do know something about first to see if they can hold solid matter together.

PERPLEXED STUDENT: But haven't you forgotten the most powerful force of all—Love?

PROFESSOR: You are certainly correct about love's power, but in this book we're interested in what holds solid matter together; and love is supposed to hold people together. Unfortunately, observation seems to indicate that solids are more together than people.

4.4

Picking the Winner

Now that we have seen what *the* force must do, and we have looked at some potential candidates for *the* force, we should be able to compare them to see which one could be the "winner." Let's be very organized and do this in Table 4.2.

Calculation #1: Let's focus our attention on the bottom atom in a crystal of sodium chloride and see if the gravitational force between this atom and its neighbor is large enough to keep the earth from ripping off the bottom atom due to its weight.

$$F_G = G\frac{m_1 m_2}{r^2}$$

$$= 7 \times 10^{-11}\,\frac{\text{nt}\cdot\text{m}^2}{\text{kg}^2}\,\frac{(4 \times 10^{-26}\ \text{kg})(6 \times 10^{-26}\ \text{kg})}{(3 \times 10^{-10}\ \text{m})^2},$$

$$= 2 \times 10^{-42}\ \text{nt}.$$

$$w = mg = 6 \times 10^{-26}\ \text{kg}\cdot 10\ \text{m/sec}^2,$$

$$= 6 \times 10^{-25}\ \text{nt}.$$

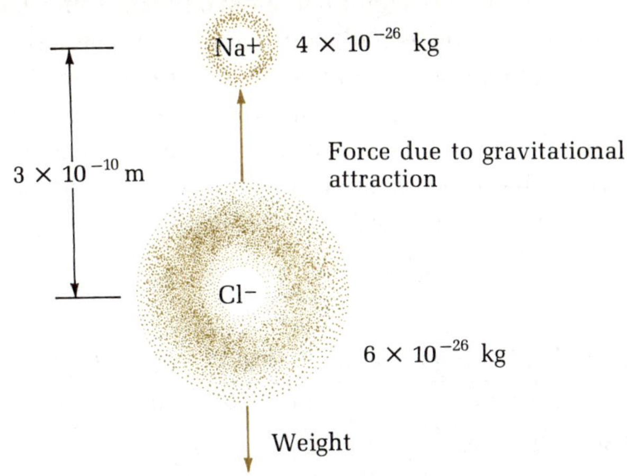

So, gravitational attraction between the sodium and chloride pulls the chloride *up* with a force of 2×10^{-42} nt, while the earth pulls the chloride *down* with a force of 6×10^{-25} nt. Who wins this miniature tug-of-war? (If you wrote out these numbers with all their zeros, it would be obvious.) The gravitational attraction between the sodium and chloride is *not* strong enough to overcome the weight of the chloride.

Calculation #2: Although we've been talking about solid

Table 4.2

Requirement → / Force ↓	Attractive and Repulsive	Greater than Weight	Universal
Gravitational	No. Attractive only.	See calculation #1.	Yes. Works on all masses.
Electric	Yes. Positive and negative charges.	See calculation #2.	Yes. (See experiment 4.2 A and B.)
Magnetic	Yes. North and south poles.	Yes. Lifts iron objects.	No. Only works on iron, nickel, and gadolinium.
Nuclear	No. Attractive until 10^{-15}m.	Yes. Nuclei don't fall apart.	No. Doesn't work on electrons, nor at greater than nuclear distances.

crystals as though they contained atoms, recall from Chapter 2 that *ions* may be formed when atoms gain or lose electrons. The sodium atom has given away its outer electron, leaving it charged +1, while the chlorine has obtained an electron, leaving it charged −1. Let's see whether the electrical force of attraction between these ions can overcome the weight of the bottom chloride ion.

$$F_E = K\frac{Q_1Q_2}{r^2},$$

$$= 9 \times 10^9 \frac{\text{nt} \cdot \text{m}^2}{(\text{coul})^2} \frac{(1.6 \times 10^{-19} \text{ coul})(1.6 \times 10^{-19} \text{ coul})}{(3 \times 10^{-10} \text{ m})^2},$$

$$= 2.56 \times 10^{-9} \text{ nt.}$$

$$w = mg = 6 \times 10^{-26} \text{ kg} \cdot 10 \text{ m/sec}^2 = 6 \times 10^{-25} \text{ nt.}$$

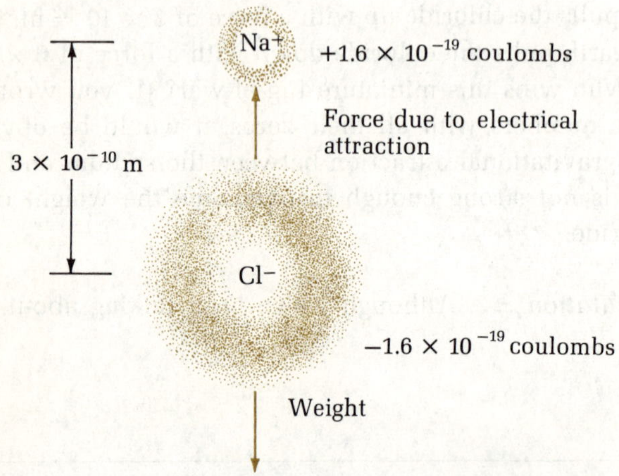

Hence, the electrical attraction of the chloride and the sodium ions yields an *upward* force of 2.56×10^{-9} nt, while the weight pulls *down* with a force of 6×10^{-25} nt. The electrical force between the ions is strong enough to overcome gravity.

The *nuclear force* fails because it works over too short distances and is not universal; the *magnetic force* is not universal; the *gravitational force* is only attractive and hopelessly weak. Only the *electric force* meets all our requirements, so

the electric force is the winner.

We've come a long way. We can now show that the electric force is the best candidate for *the* force which holds ions in their places in solid matter, but there remain quite a number of details to be sorted out. For example, it's easy to see how the attractive force works to hold ions *together*, but what keeps these charged objects *apart*? Or, what (if any) role does the interesting induced-charge phenomenon play? How can other kinds of solid matter besides this ionic crystal of NaCl be explained? In short, we have acquired a degree of knowledge so far, but we still have to apply it to form a good explanation of all solid matter.

Summary

The question to be answered was: What holds the atoms in place in solid matter? The answer involved force, so we began by studying Newton's first law which says that a body in straight-line motion tends to stay in straight-line motion until acted upon by an outside force. An outside force causes a body to accelerate according to Newton's second law: $F = ma$. The gravitational force exerted by the earth on everything does *not* cause the bottom ions in a crystal to fall down; so we concluded there must be a force, strong enough to overcome the bottom ion's weight, pulling *up* on the bottom atom. Further, we decided that this bonding force must be both attractive and repulsive because solids cannot be pulled apart or squeezed together easily. Furthermore, we decided it would be simplest if just *one* force did all this bonding.

Next, we investigated several forces, beginning with the gravitational force, an attractive force that exists between any two bodies that have mass. Then we looked at the electric force that exists between any two charges, the magnetic force, the nuclear force, and the weak force. By comparing the forces with the requirements, we selected the electric force as *the* force which holds solid matter together.

Questions

This may be used like a programmed text by referring to the indicated questions while reading the text. (Note: One of the numerical answers given is incorrect. Perhaps you can find it.)

4.1

1. Define inertia and give some examples of inertia in fields other than physical science.
2. State Newton's first law and give examples from your own experience.
3. Define velocity.
4. What is the difference between velocity and speed?
5. A bug crawls across a 4 meter wall in 200 seconds. What is its speed? (0.02 m/sec)
6. If your automobile is traveling at 30 m/sec (about 70 mph), how far will it go in 0.5 sec (a normal reaction time)? (15 meters)
7. How long does it take an astronaut going 800 m/sec to travel 2 meters? (2.5×10^{-4} sec)
8. Define the term acceleration, including its units.
9. Explain where the sec^2 comes from in the units for acceleration.
10. As a small airplane roars down the runway, it accelerates from zero to 60 m/sec in about 15 sec. What is its acceleration? (4 m/sec^2)
11. What is the change in velocity for a fly who accelerates at the rate of 20 m/sec^2 for 0.5 second? (10 m/sec)
12. What is the usual intuitive definition of a force?
13. What is Newton's second law, relating force to mass and acceleration?
14. What are the units of force?
15. What is meant by net force? (Draw a diagram if necessary.)
16. How much force would be required to accelerate a 0.02 kg bug at a rate of 100 m/sec^2? (2 newtons)
17. How much acceleration would a housemover achieve if he applied a 3×10^3 newton force to a 1.5×10^5 kg house? (2×10^{-2} m/sec^2, assuming no friction)
18. What is the difference between mass and weight?
19. What is the weight of a person whose mass is 80 kg? (800 newtons)
20. What is the weight of a sodium ion whose mass is 4×10^{-26} kg? (4×10^{-25} newtons)

4.2

21. On what physical basis was it suggested that the force which holds solid matter together is both attractive and repulsive?
22. For what physical reason must the bonding force be greater than the weight of an atom?

23. Why would we like the bonding force to be universal?

4.3

24. Using Newton's law of gravitation, what is the force between two 0.020 kilogram cookies whose centers are 0.070 meters apart? (5.7×10^{-12} newtons)
25. What is the gravitational force between an electron (mass = 9×10^{-31} kg) and a proton (mass = 1.7×10^{-27} kg) about an x-ray wavelength (7×10^{-11} m) apart? (2.2×10^{-47} newtons)
26. What new unit was added to the MKS system to take care of the electric charge?
27. Using Coulomb's law, determine the electric force exerted by an electron ($Q = -1.6 \times 10^{-19}$ coul) on a proton ($Q = +1.6 \times 10^{-19}$ coul) located 3×10^{-10} meters away. (2.6×10^{-9} newtons)
28. How much electrical force is exerted by a +2 coulomb charge repelling a +8 coulomb charge at a distance of 200 meters? (3.6×10^{6} newtons)
29. Explain how the magnetic force works.
30. Explain how the nuclear force works.

Quotes to Ponder

1. "Weight watchers would be out of luck in space because bathroom scales, or any other scale for that matter, won't work where there is no gravity. The best 'honest weight, no springs' scales would show zero pounds because the person being weighed would be weightless.

However, this is a problem the National Aeronautics and Space Administration (NASA) has found necessary to solve because experiments in the space agency's upcoming Skylab program involve weight checks of the astronauts to determine what happens to their bodies during long periods of weightlessness.

NASA scientists have solved the problem of determining the mass of crew members who will work in the space laboratory for up to 56 days and laboratory materials with three mass measurement devices. The devices operate independent of gravity.

Specimens to be weighed are placed in pans, or in the case of the astronauts in a chair. The pan or the chair is set in motion by springs which exert a known force (F) on the subject. The acceleration (A) of the moving subject is determined by a photo-optical device which measures motion.

With known values for force and acceleration, it is a simple matter to calculate the mass (M) by using the formula students use in basic physics courses—F equals ma." ("The Science Beat" by Larry Bush, *The Ann Arbor News*, October 19, 1971)

2. "Speed limits now are posted in the kilometers as well as miles per hour in the Rocket City.

Some 268 of the metric signs have been erected on major thoroughfares in Huntsville, where the Apollo moon rocket was born.

For instance, a 40 miles per hour sign has a 65 kilometers per hour sign below it.

The signs were provided by the metrication committee at Marshall Space Flight Center in an effort to make people aware of the metric system.

The committee is an informal group of workers at the National Aeronautics and Space Administration installation. (AP News Release, Dateline Huntsville, Alabama, May, 1971)

CHAPTER FIVE

PUTTING TOGETHER THE VARIOUS KINDS OF SOLID MATTER

We have now seen that ionic crystals consist of positive and negative ions arranged in a regular fashion with their centers about an x-ray wavelength (10^{-10} meters) apart and held in place by the electric force. In this chapter, we will formulate a simple conceptual model for ionic crystals, discover shortcomings in our model, and investigate the concept of energy to take care of the difficulties. Also, we will formulate a more complicated ionic crystal model with energies in it, and then we'll investigate other forms of solids, namely molecular crystals, molecular noncrystals, and other solids, to see how our conceptual model may be modified to explain the characteristics of solid matter.

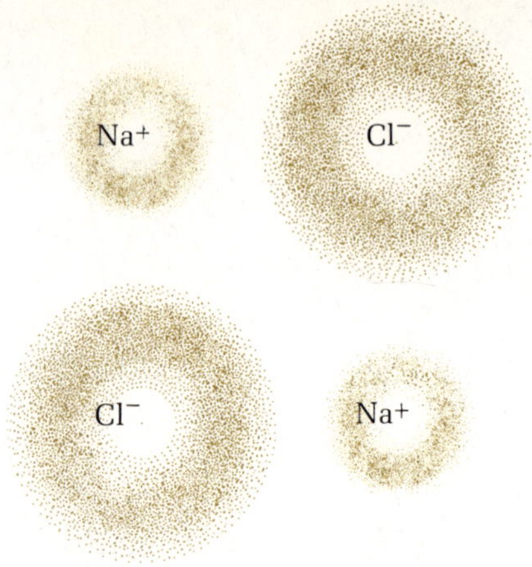

Figure 5.1 *Diagram of the static ionic crystal conceptual model*

5.1

Static Ionic Crystal Conceptual Model

We can now formulate our first conceptual model of an ionic crystal. A crystal of our old friend sodium chloride consists of a regularly arranged series of sodium and chloride ions with their centers about an x-ray wavelength apart (see Figure 5.1).

Electric attraction of the Cl^- and the Na^+ keeps the crystal from coming apart. Recalling the nature of ions from Chapter 2, the "fuzzy" outer surfaces on the diagram in Figure 5.1 represent the outer fringes of the electron cloud, rather than some solid surface. Thus, if the Na^+ and Cl^- get too close together, do these "clouds" touch and prevent the ions from coming closer together, or can they overlap (Figure 5.2)?

The electron clouds are *not* solid surfaces, so these ions *may* overlap. If the ions do overlap, these electron clouds are both negative, so a repulsive force is developed, preventing further overlap.

Since the ions are not moving, this conceptual model will be called the *static ionic crystal conceptual model*.

This model is a bit too simple, as we shall discover in the next experiment.

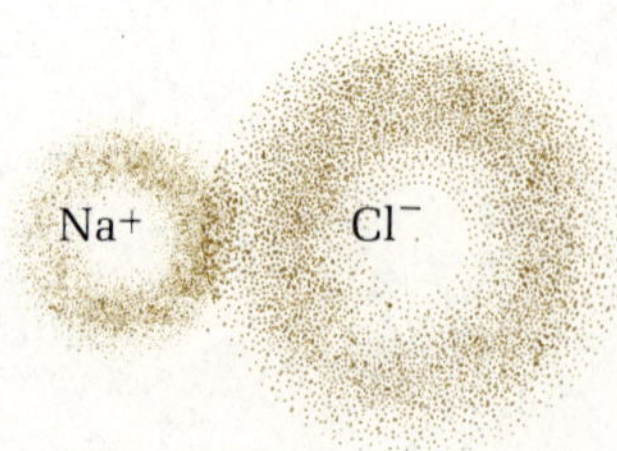

Figure 5.2 *Can the ions in an ionic crystal overlap?*

EXPERIMENT 5.1

IONIC CRYSTAL COOLING CURVE

In this experiment, we are going to melt a solid crystalline substance and measure the temperature of this substance as it cools, changing from a liquid to a solid (see Figure 5.3). (The substance is not truly an ionic crystal, but a molecular crystal which acts just like an ionic crystal. As we'll see later, ionic crystals melt at extremely high temperatures, and we'd be here all day waiting for the thing to cool down, not to mention fire hazards.)

Procedure

1. *Wear your safety glasses and have a fire extinguisher nearby.*

2. *Figure out how the heating device works. (If it is a bunsen burner, fiddle with the gas and air controls until you get a flame with a light blue cone in the center. The hottest part of the flame is just above this cone.)*

3. *Melt the substance in the test tube by playing the flame all around the bottom of the test tube. Don't get the flame near the top or you may ignite the fumes. Make sure the substance is completely melted.*

4. *Insert the thermometer into the test tube and begin taking temperature readings every 30 seconds, starting when the crystals inside the test tube are molten and above 100°C* until it is completely solid and at about 40°C. (Room temperature is about 22°C, but it will take it quite a while to get there.)*

(Hint: Position the thermometer so that you can read it without moving it or taking it out of the tube. This will eliminate both inaccurate readings when the substance is a hot liquid and a broken thermometer when it is solid. To get the thermometer out, reheat the substance enough to liquefy it.)

5. *Plot your temperature-time results in the form of a graph. It will probably look something like the one illustrated in Figure 5.4.*

6. *This is disturbing! Why does it hesitate near 80°C for all that time? Remelt the substance, let it cool again, and lightly touch the outside of the test tube occasionally to see if it is still giving off heat, particularly when it is "stuck" near*

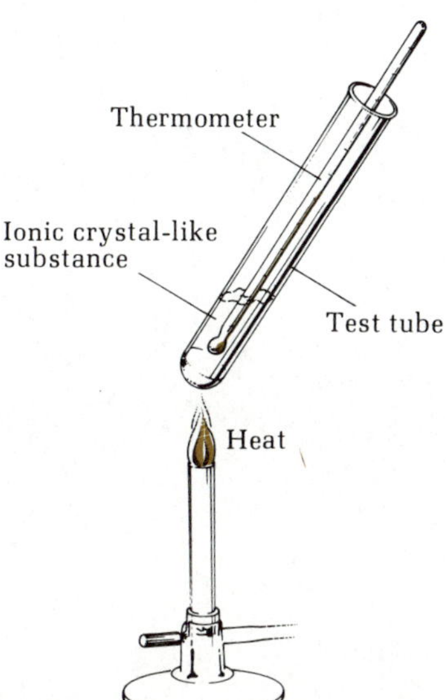

Figure 5.3 *Apparatus for determining the cooling curve of an ionic crystal*

*°C stands for degrees celsius, a temperature scale in which water freezes at 0°C and boils at 100°C, unlike our fahrenheit scale (°F) in which water freezes at 32°F and boils at 212°F. The celsius scale is often used in scientific work.

80°C. CAUTION: It is still hot, so touch it very lightly.*

The big flaw in our static ionic crystal conceptual model is that it doesn't deal at all with temperature, so it couldn't possibly explain why the temperature remains constant while the substance changes from a liquid to a solid. Before we can modify our conceptual model, we'll have to take a small side trip to look at the concept of energy.

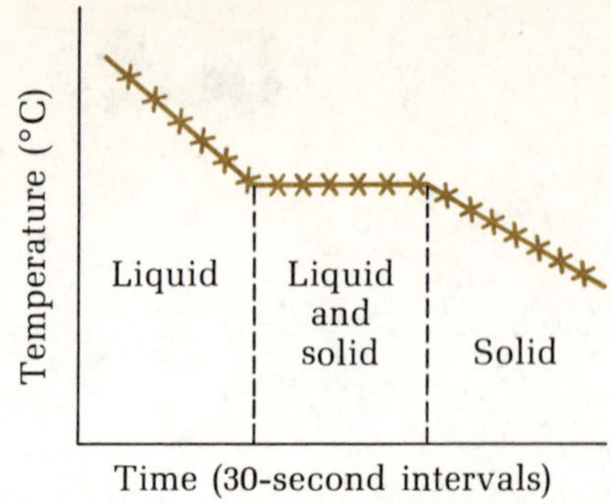

Figure 5.4 *Ionic crystal cooling curve*

5.2

Energy and Its Conservation

We'll have to build some basic concepts about energy before we can apply energy to ionic crystals. Let's start by putting ourselves into some situations where our intuition will tell us some things about energy.

Situation #1: You are a goalie in a fast-moving hockey game (Figure 5.5). Do you feel a different amount of "sting" in your mitt if you stop a very rapid slap shot from the point, as opposed to a slow-moving backhand dribbler? Sure you would. If we can believe that the "sting" is related to the energy of the puck, then our intuition tells us that *as the speed of a body increases, so must the body's energy.*

Situation #2: You are a catcher in a baseball game (Figure 5.6). A pitcher with a weird sense of humor throws a stone of the same size and shape as a baseball. Do you notice any difference in the "sting" in your mitt? Again, relating the "sting" with the energy, our intuition says that *as the mass of a body increases, so must the body's energy.*

Situation #3: You are a fireman, about to catch a person jumping out of a burning building (Figure 5.7). Do you think you'd notice much difference in the deflection of your net if the person jumped from the second or from the fourth floor? Relating the deflection of the net to the energy, our intuition says that *as the height of a body increases, so must the body's energy.*

Now that we have our intuition all squared away, let's see how well the quantitative formulation of energy matches our ideas.

There are basically *two* kinds of energy, *kinetic energy* and *potential energy.*

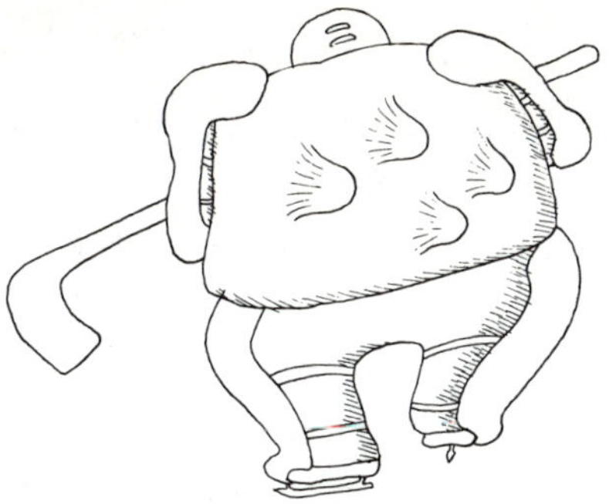

Figure 5.5 *You as a hockey goalie*

Figure 5.6 *You as a baseball catcher*

Figure 5.7 *You as a fireman*

The word kinetic means *motion,* so kinetic energy is energy possessed by a body in motion.

$$\text{Kinetic Energy} = KE = \frac{1}{2}mv^2,$$

where m = mass of the body (kg),
v = speed of the body (m/sec).
Note that a body at rest ($v = 0$) has no kinetic energy.

Potential energy is energy "stored up" due to a body's *position.* It may come in various forms, such as gravitational potential energy, elastic potential energy, electrical potential energy, and others. For gravitational potential energy,

$$\text{Gravitational Potential Energy} = PE = mgh,$$

where m = mass of the body (kg),
g = acceleration due to gravity (≈ 10 m/sec^2),
h = distance above some reference level (meters).

Let's work with gravitational potential energy first, and introduce the other forms of potential energy later.

Using Newton's second law and the definitions of weight, acceleration, and velocity from Chapter 4, it may be shown that, in the absence of friction, the sum of the kinetic energy and the potential energy of an isolated system of bodies is constant. This is the *principle of conservation of total energy.*

Total Energy = Kinetic Energy + Potential Energy.
$TE = KE + PE$ = Constant for an Isolated System.

An isolated system is simply a system which exchanges no energy with any other system. For example, a perfectly insulated box or room would be an isolated system.

As a check, let's make sure both kinds of energy have the same unit.

$$KE = \frac{1}{2}mv^2 = [\text{kg}]\left[\frac{\text{m}}{\text{sec}}\right]^2 = \frac{\text{kgm}^2}{\text{sec}^2}.$$

$$\text{Gravitational } PE = mgh = [\text{kg}]\left[\frac{\text{m}}{\text{sec}^2}\right][\text{m}] = \frac{\text{kgm}^2}{\text{sec}^2}.$$

This particular combination of fundamental units is given a special name in honor of a scientist (Sir James Joule) who originally worked with energy:

$$\frac{\text{kgm}^2}{\text{sec}^2} = \text{joule (pronounced "jewel").}$$

Notice that your intuition was right:

KE increases as v increases,

Gravitational PE increases as h increases,

Both KE and PE increase as m increases.

Let's do some example problems so that energy becomes a good friend rather than just an acquaintance.

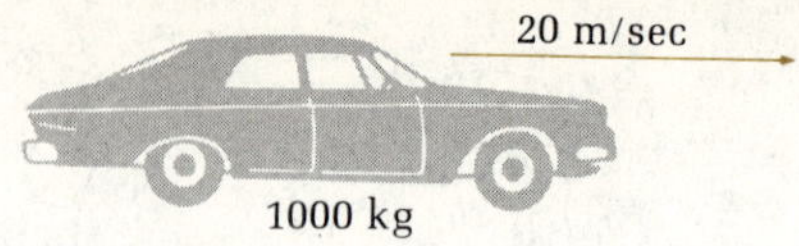

Figure 5.8 *Automobile with KE*

Example #1: What is the kinetic energy of a 1000 kg automobile going at 20 m/sec (Figure 5.8)? (Fasten your seatbelts.)

$$KE = \frac{1}{2}mv^2 = \frac{1}{2}(1000 \text{ kg})(20 \text{ m/sec})^2 = 2 \times 10^5 \text{ joules}.$$

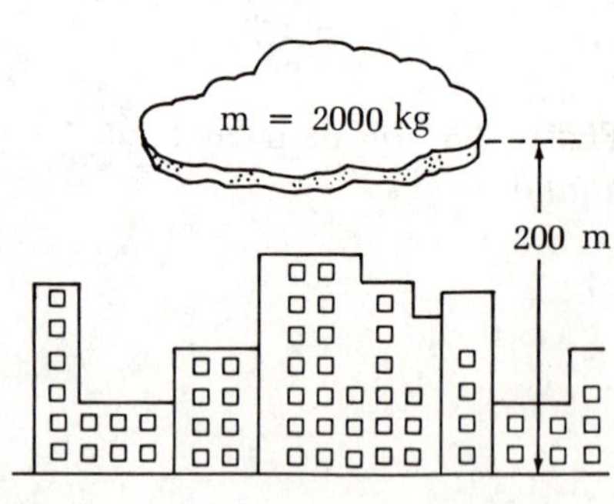

Figure 5.9 *Smog cloud with gravitational PE*

Example #2: What is the gravitational potential energy of a 2000 kg smog cloud which is sitting motionless 200 meters above the city, waiting to descend (Figure 5.9)?

$$\text{Gravitational } PE = mgh = 2000 \text{ kg} \cdot 10 \text{ m/sec}^2 \cdot 200 \text{ m}, \quad = 4 \times 10^6 \text{ joules}.$$

Example #3: What is the total energy of a 60 kg skier going 10 m/sec along a bluff which is 20 m above the flat plain (Figure 5.10)?

$$TE = KE + PE = \frac{1}{2}mv^2 + mgh,$$

$$= \frac{1}{2}(60 \text{ kg})(10 \text{ m/sec})^2 + 60 \text{ kg}(10 \text{ m/sec}^2)(20 \text{ m}),$$

$$= 3000 \text{ joules} + 12{,}000 \text{ joules},$$

$$= 15{,}000 \text{ joules}.$$

If the skier jumped off the bluff, what would be the total energy of the skier just before he hit the plain? 15,000 joules (conservation of total energy).

Let's do an experiment and see if we can verify the law of conservation of total energy.

EXPERIMENT 5.2
ENERGY CONSERVATION

Galileo is said to have spent a good deal of his time in church observing the swing of a light hung on a long chain from the church ceiling. (Maybe that helps explain why Galileo was hassled so much by the church.) A pendulum

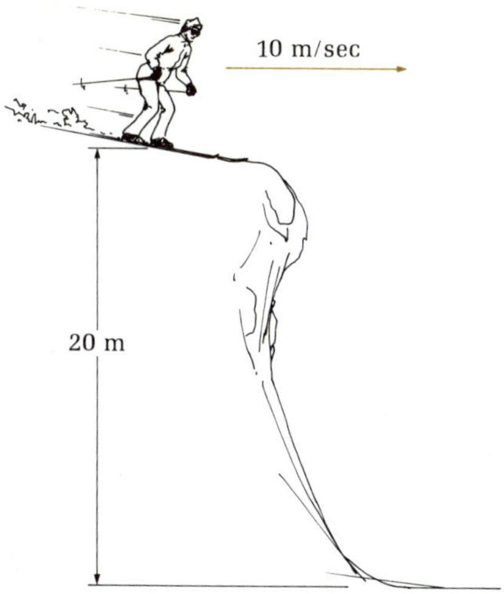

Figure 5.10 *Skier with KE and gravitational PE*

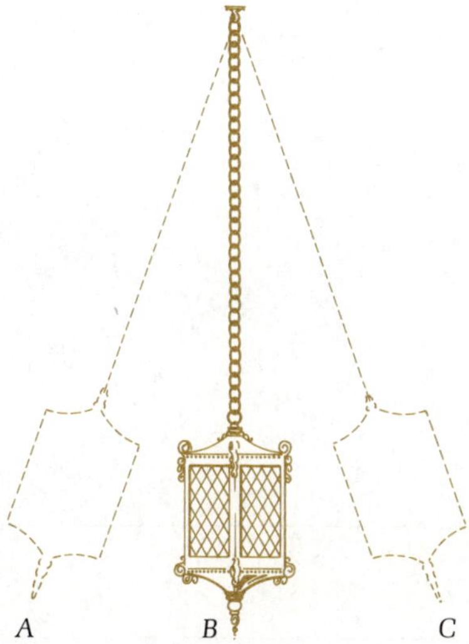

Figure 5.11 *Light on a chain acting as a pendulum*

does have some very interesting properties in terms of its energy, so let's have a close look at a pendulum (Figure 5.11).

At what position does the pendulum bob have its maximum gravitational potential energy? Since gravitational PE depends on the height, the maximum gravitational PE will occur at the maximum height of the pendulum bob, which happens at the ends of the swing, either point A or C in Figure 5.11. In the middle of the swing, at point B, the pendulum is at its lowest height; so it should have a minimum gravitational PE at point B.

At what point does the pendulum bob have its maximum kinetic energy? Since kinetic energy is energy of motion, depending on v^2, the maximum KE must occur at the point where the bob is at its maximum speed. This must be at point B in Figure 5.11, the lowest point in the swing, since it slows down going up either side. At the ends of the swing, the pendulum has stopped; then it turns around and comes back, so the KE at the ends must be zero because there is no velocity (Figure 5.12).

If we had some way of measuring h and v, we could verify the law of energy conservation. We can measure h by using a meter stick and v by using the old markable tape and dot-maker trick we used to measure g (Experiment 4.1).

Procedure

1. *Set up the pendulum bob with the markable tape attached as shown in Figure 5.13, and measure the height of the bob above any convenient reference level (e.g., floor, table) when the string is vertical. This is called h_1 in Figure 5.13.*

2. *Pull the pendulum bob up to position 2 and measure h_2, the distance above the reference level, threading the markable tape into the dot-maker.*

3. *Turn on the dot-maker and release the pendulum bob from position 2. It will move downward, dragging the markable tape through the dot-maker. The tape it generates should look like that illustrated in Figure 5.14.*

4. *Using the timing information supplied, determine the velocities over each time interval between ticks of the dot-maker, recalling that*

$$v = \frac{\text{distance}}{\text{time}}.$$

5. *Presuming that the velocity at the bottom of the swing* *is the maximum velocity, see if energy conservation holds:*

$$\frac{1}{2}mv_2^2 + mgh_2 \stackrel{?}{=} \frac{1}{2}mv_1^2 + mgh_1.$$

(Does $v_2 = 0$?)

6. *The two sides of this equation will undoubtedly not be precisely equal. Calculate the percent difference.*

$$\text{Percent difference} = \frac{\text{one side} - \text{other side}}{\text{either side}} \times 100.$$

Explain anything in the experiment or measuring process which you think might cause this difference. The point is to convince yourself that energy is really conserved.

7. *Using the principle of energy conservation and your tape, calculate the potential energy over each time interval for several time intervals on either side of position 1.*

$$PE = TE - KE,$$
$$= TE - \frac{1}{2}mv^2,$$
$$= mgh_1 + \frac{1}{2}mv_1^2 - \frac{1}{2}mv^2.$$

8. *Plot the potential energy vs. time on a graph. It should look something like Figure 5.15.*

This curve is referred to as a potential energy well, and we will see it many more times.

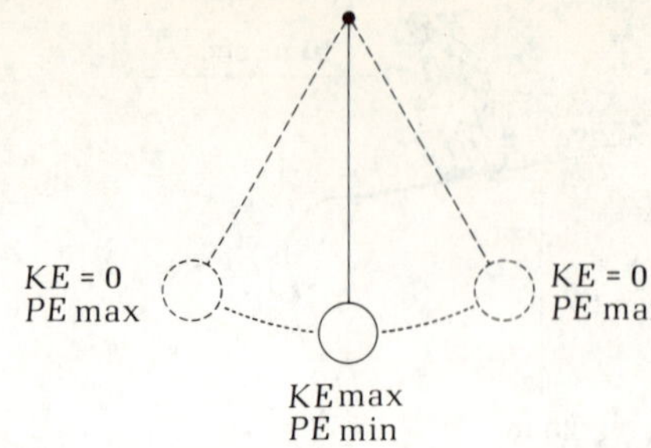

Figure 5.12 *Pendulum showing energy variations*

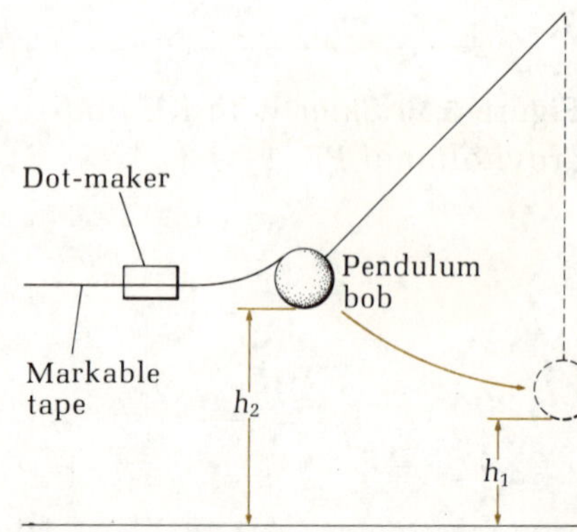

Figure 5.13 *Pendulum experiment setup*

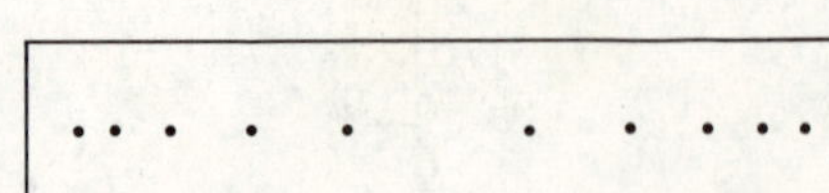

Figure 5.14 *Tape generated by pendulum*

So far, we have discussed only one kind of potential energy—gravitational potential energy. Gravitational potential energy is stored up by "lifting" a body against the gravitational force of attraction (Figure 5.16).

Although we normally talk about lifting a body "up" to increase its gravitational potential energy, "up" is a relative term. As Figure 5.16 shows, a body at position 2 may be considered "up" with respect to position 1 only in the sense that it is farther from the center of the earth. Thus, it is really the action of pulling against the attractive force which builds up the potential energy.

Let's see if we can generate some potential energy using a force other than the gravitational force. Envision a body sitting on an almost frictionless surface, like a skater on very slick ice, as shown in Figure 5.17. Now attach two springs, acting in opposite directions, to this body.

If you were to move the body in either direction, you

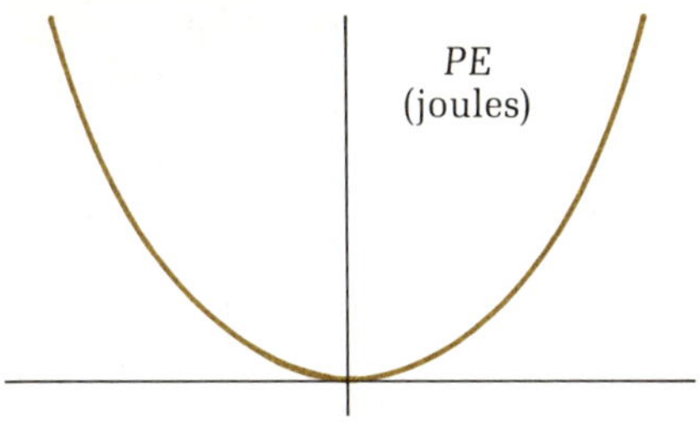

Figure 5.15 *Pendulum potential energy well*

Figure 5.16 *Storing gravitational potential energy by moving a body "up"*

Figure 5.17 *Body held by two springs on a low-friction surface*

would stretch the spring and build up *elastic potential energy*. Once you let it go, the body would move in the other direction, building up kinetic energy as it increased its speed, reaching a maximum speed at some point, then going too far and stretching the other spring until the body finally stops and turns around, going back the other way.

If we drew a graph of the potential energy of the body as it moved back and forth, it would look like Figure 5.18.

The total energy is a constant according to the principle of conservation of energy, so it is plotted as a horizontal line. The kinetic energy at any point is just enough to make up the difference between the total energy and the potential energy. Notice that the *KE* is a maximum in the center and decreases to zero at each end. This whole picture should look familiar, since we drew the same sort of graph in Experiment 5.2 for the pendulum. The only real difference, from the energy standpoint, is that the pendulum used gravitational potential energy, while this body uses elastic potential energy.

5.3

Kinetic Ionic Crystal Conceptual Model

Now comes the big imaginative leap: How about *electrical potential energy*? Suppose we had a negative ion and a positive ion, as shown in Figure 5.19.

As we have seen previously, when the positive ion moves far enough toward the negative ion, repulsive forces develop as the electron clouds overlap, and the positive ion is shoved in the other direction, until there is no overlap, and the attraction between the positive and negative takes over, bringing it back. The situation is completely analogous to the pendulum and the body between the springs, so the potential energy diagram (Figure 5.20) is also similar.

SHARP STUDENT: Oh, I see. In the other two instances, potential energy was stored up by working against gravity and against a spring restoring force. In this case, with ions in a solid, the potential energy is built up against the electrical force of attraction at long range and repulsion at short range. How does this work when you have more than just these two ions?

PROFESSOR: In an ionic crystal where there is a regular array of ions, the potential energy wells are extremely regular also (see Figure 5.21).

STUDENT: *Then the ions are not really stationary.* They are actually bouncing around like a pendulum or like a skater between two springs. They don't really get "away," however, because their total energy is not enough for them to escape from their wells; they just bounce around inside. Is there any way we can sense all this jostling around?

PROFESSOR: Good question. Suppose we had a device that would expand when it was jostled around by these moving ions. We could judge the activity of the ions, and hence their kinetic energy, by the amount of expansion of our device. Such a device does exist; it's called a *thermometer* (Figure 5.22).

STUDENT: What you're saying, then, is that what we call temperature is actually related to the kinetic energy of the individual ions that bang into the thermometer. When the *KE* goes up, the temperature goes up, and vice versa.

PROFESSOR: The transfer of kinetic energy from one body to another is accomplished quite simply. Envision two boxers, one of whom is very active, dancing around, throwing lots of hard punches (lots of *KE*), while the other doesn't move much and throws occasional weak punches (not much *KE*). Who usually wins? In the same way, the ion with the high *KE* almost invariably shares its *KE* with lesser *KE* ions by "colliding" with them. (They don't collide in the same sense that you think of solid bodies colliding; all they need to do is get their electron clouds overlapping and they'll be strongly affected.)

STUDENT: I'd say that's cool, but you'd only accuse me of making a bad pun. If I accept the regular potential well structure of ionic solids and the idea that the temperature is proportional to the average kinetic energy, what would an energy diagram for a liquid look like?

PROFESSOR: Let's discuss this question and see what we can come up with. *First* of all, liquids generally flow and take the shape of their containers; so *we shouldn't expect* an orderly array of ions, nor an orderly potential well structure. *Secondly,* liquids generally have a smaller density than their corresponding solid, as evidenced by the solid sinking to the bottom of the liquid in Experiment 5.1.

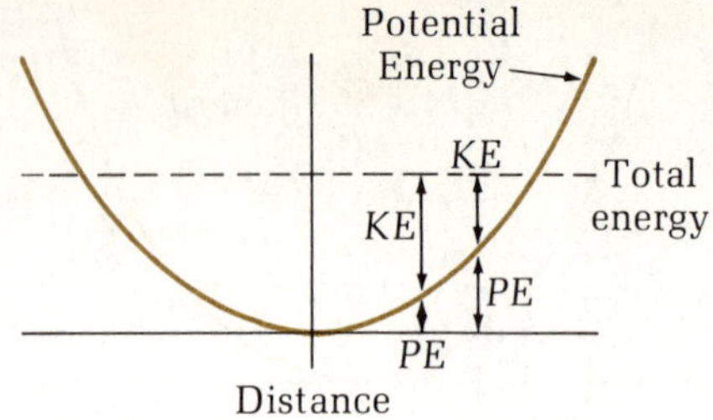

Figure 5.18 *Potential energy vs. distance for a body moving under the influence of two springs*

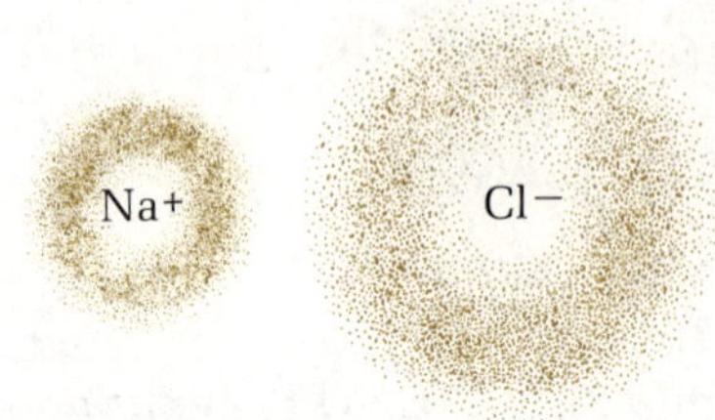

Figure 5.19 *Positive ion and negative ion*

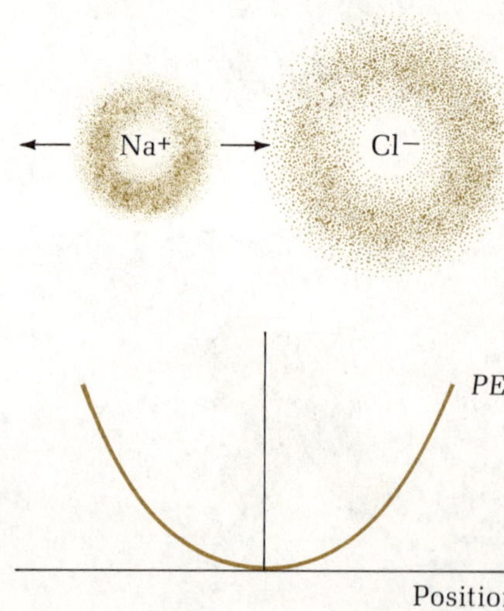

Figure 5.20 *Potential energy diagram for a positive ion*

Figure 5.21 *Potential energy wells in an ionic solid*

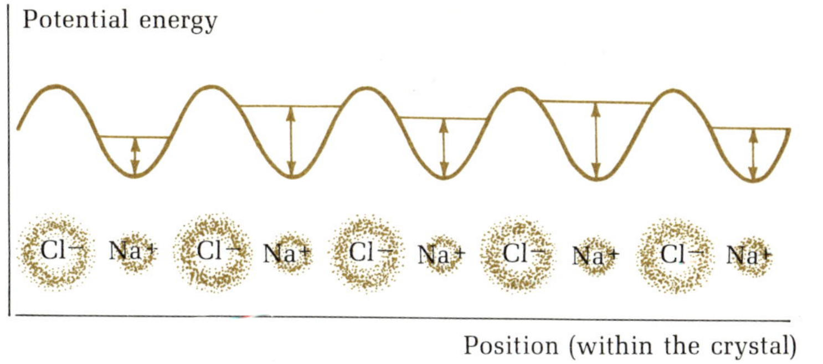

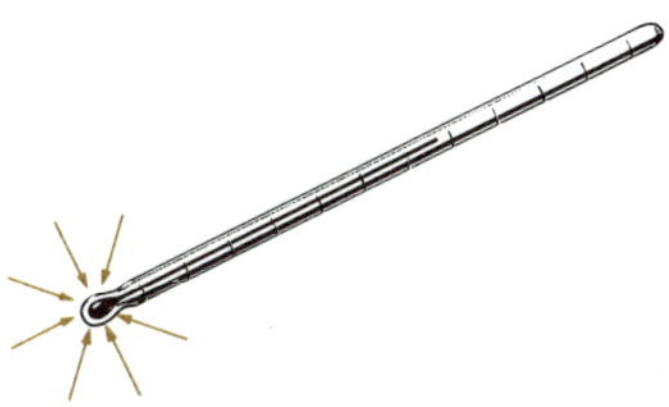

Figure 5.22 *Ions colliding with mercury in a thermometer, sending it soaring*

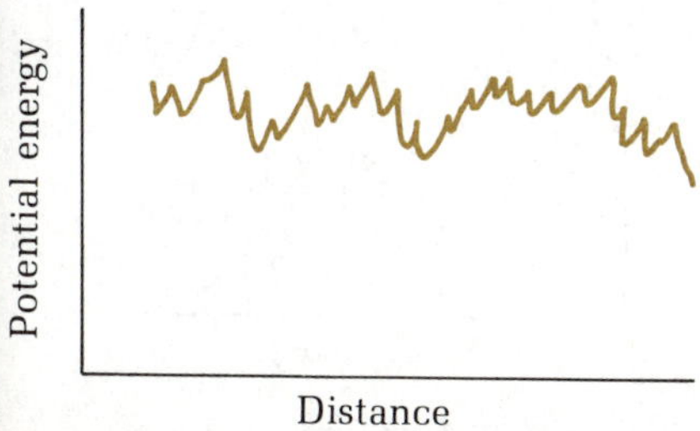

Figure 5.23 *Potential energy diagram for ions in a liquid*

(Water is an important exception to this rule, with ice floating in liquid water. Water is such a special case, we'll devote all of Chapter 8 to it.) Now, if the density of the liquid is lower than the density of the solid, then the average distance between ions in the liquid must be greater than the average distance between ions in the solid. (Think about it.) To pry the ions farther apart implies greater potential energy, so the *average potential energy in the liquid must be greater than in the solid.*

A potential energy diagram for a liquid must look like Figure 5.23.

STUDENT: In the liquid, then, ions are relatively free to come and go as they please, because there are no deep wells to "trap" them. On the other hand, their average potential energy must be higher since the ions are farther apart in the liquid than in the solid. I think I'm beginning to catch on to what happened in Experiment 5.1.

PROFESSOR: Good! To understand that experiment fully, let's first revise our hypothetical model for ionic crystals, then apply it to all regions of the cooling curve in Experiment 5.1.

An ionic crystal consists of positive and negative ions arranged in a regular array (Figure 5.24). The ions are in constant motion, interchanging potential and kinetic energy similar to the way a pendulum does. On the average, the ions are about an x-ray wavelength apart, and it is the electric force that keeps them moving and interchanging *KE* and *PE*. The ions have different amounts of *KE*, but their average *KE* may be measured, and the temperature is proportional to the average *KE*.

The liquid obtained when the ionic crystal is melted differs from the solid in that there is no regularity to the liquid structure. The liquid potential energy wells are

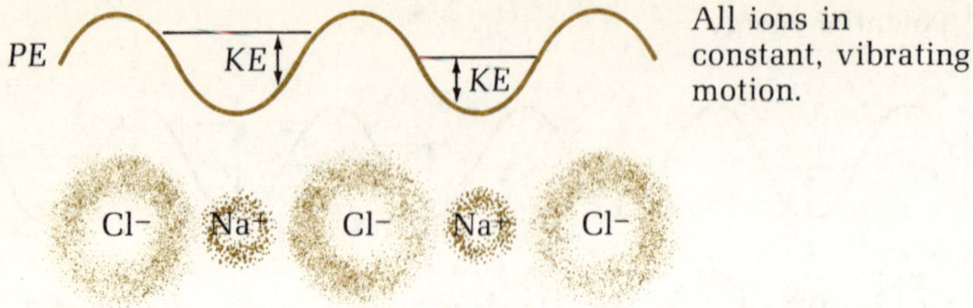

Figure 5.24 *A conceptual model of a kinetic ionic crystal*

irregular and more shallow than the solid potential energy wells, and the average potential energy in the liquid is greater than that of the solid because the ions are farther apart.

Now, let's try to explain the cooling curve we obtained in Experiment 5.1. Recall, it looked like the one illustrated in Figure 5.25.

In region A, only liquid is present; so as the temperature decreases, the average kinetic energy of the ions in the liquid decreases, and the liquid becomes less "active" as the ions move about less wildly. If you touched the test tube in this region, you would find it hot to the touch, meaning that your fingers or the surrounding air is increasing its temperature and hence its kinetic energy.

Let's consider the laboratory to be an isolated system so that we can apply the law of conservation of total energy to the bodies within the laboratory. We can subdivide the contents of the laboratory into two parts: (1) the ionic crystals inside the test tube (in either the solid or liquid state), and (2) the ionic crystals' surroundings (e.g., the test tube, the air in the room, and your finger when you touch it). Now, if this is truly an isolated system, conservation of total energy requires that (*KE* + *PE*) *of the ionic crystals* + (*KE* + *PE*) *of the surroundings must remain constant.* In region *A*, the liquid is cooling and its temperature is decreasing, so *KE* of the liquid ionic crystals is decreasing. The test tube, the surrounding air, and your finger when you touched the test tube are all increasing their temperature and hence the *KE* of the surroundings is increasing. Total energy is conserved because the *KE* decrease of the liquid ionic crystals is balanced by the *KE* increase of the surroundings.

In region C, only solid is present, and the situation is quite similar to that of region *A*. The kinetic energy of the ions in the solid decreases as the kinetic energy of

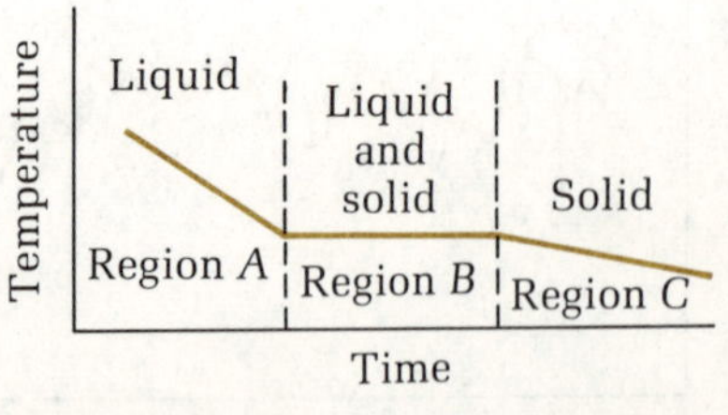

Figure 5.25 *Cooling curve obtained from Experiment 5.1*

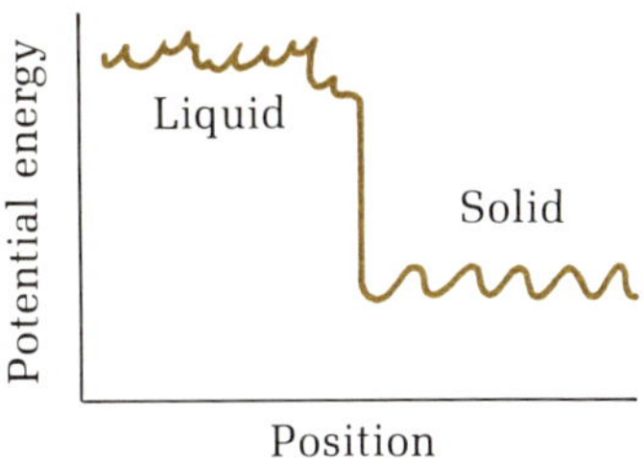

Figure 5.26 *Potential energy diagram for solid and liquid together*

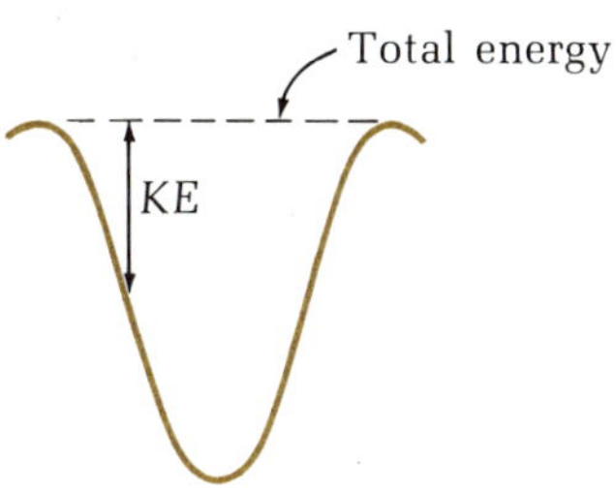

Figure 5.27 *Ion in potential energy well at melting temperature*

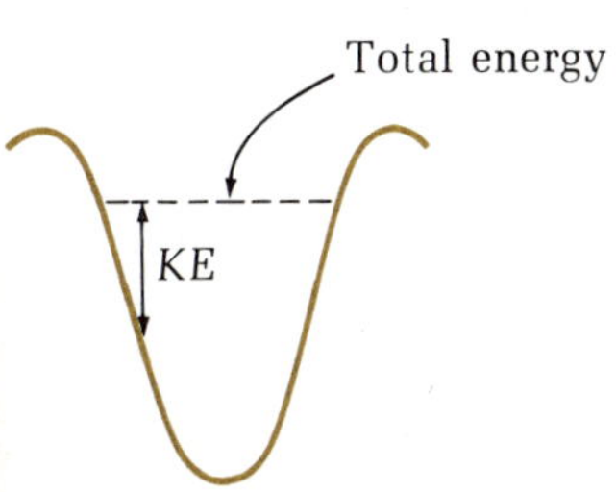

Figure 5.28 *Ion in potential energy well below melting temperature*

the surroundings increases. The big difference between region *C* and region *A* is that the ions are considerably closer together in the solid region *C*.

In region B, the real fun begins. Both the liquid and solid are present, so let's have a look at a potential energy diagram in which both liquid and solid exist together (Figure 5.26). (Note: This occurs at $\approx$80°C.)

During this time period, the average kinetic energy of the ionic crystals is constant, as evidenced by the constant temperature. Yet, as you discovered by touching the test tube, kinetic energy is still being transferred to the surroundings. Where does this kinetic energy come from?

SHARP STUDENT: I think I understand. As the ions go from the far-apart high average potential energy state in the liquid to the close-together low average potential energy state in the solid, they must lose the difference in potential energy somewhere, and it shows up as increased kinetic energy of the surroundings. But, why doesn't the kinetic energy of the ions change during this process?

PROFESSOR: As your observations indicated, the temperature and hence the average kinetic energy of the ions in both the solid and liquid remained constant while the solid was being formed. The kinetic energy doesn't change because it is just the right amount for the ion to fit into the potential well formed by the other ions (see Figure 5.27).

This would be analogous to the pendulum going back and forth from horizontal to horizontal or the skater going from almost breaking the spring in one direction to almost breaking the spring in the other direction.

Once the solid is formed, the temperature and hence the *KE* drops, as shown in Figure 5.28.

This implies that the melting temperature is governed by well "depth," which is in turn determined by the strength of the bond between ions. Your intuition had already told you that the stronger the bond, the higher the melting temperature, so the conceptual model simply confirms what you already knew. Ionic bonds are quite strong, so the potential energy wells associated with ionic crystals are quite deep, and melting temperatures of ionic crystals are consequently quite high, typically 200°C or more.

The substance we melted in Experiment 5.1 was actually *not* an ionic crystal, which is why it melted as low as 80°C. This bit of trickery was done simply because of safety, and the essential features of the cooling curve were the same as if the substance had been an ionic crystal. To summarize the point, let's look at Figure 5.29.

STUDENT: In essence, then, when a substance is an ionic liquid or an ionic crystalline solid, it cools off by transferring kinetic energy to its surroundings. But, when it changes from a liquid to a solid, its kinetic energy remains constant, and the change in potential energy caused by the ions moving closer together to form a solid shows up as increased kinetic energy of the surroundings. Wow!

PLEASED PROFESSOR: That's right; and now that you've mastered ionic crystals, let's see what we can do with other kinds of solid matter.

5.4

Molecular Crystals

Once we got it all sorted out, it wasn't too difficult to understand how crystals are formed from ions. After all, ions are charged positively and negatively, and it seems pretty natural to fit them together to form a solid.

Unlike ions, *molecules have no net charge,* so you can imagine it's going to be a bit tricky to hold them together with the electric force.

First of all, we'd better figure out what molecules are before we try sticking them together. Instead of physically transferring electrons from one atom to another to form ions, suppose there were atoms which do not *completely give up* or *completely take* electrons, but instead *share* electrons in order to obtain the lower energy associated with filled outer shells. This idea of sharing a pair of electrons between two atoms is the basis for what is called the *covalent bond.*

As an example, let's look at the water molecule, H_2O. Oxygen has an atomic number of 8, so it has 8 electrons, arranged in a 2 + 6 fashion. If it had just two more electrons in the second energy level, it would have a filled outer shell and become an oxygen ion. We might suppose it would take one electron from each of two hydrogens, leaving two hydrogen ions with no electrons. Not so! The oxygen atom cannot take either hydrogen's electron away completely, so it turns out that the *oxygen and two hydrogens each share*

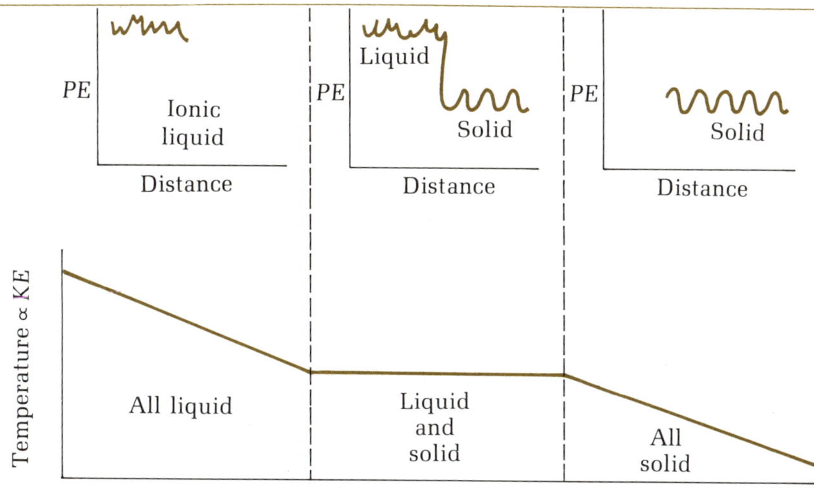

					Source
Ionic crystal	PE	Constant	Decreasing	Constant	From conceptual model
	KE	Decreasing	Constant	Decreasing	From measured temperature
Surroundings	PE	Constant	Constant	Constant	
	KE	Increasing	Increasing	Increasing	From touching test tube

Figure 5.29 *Cooling curve energy summary*

a pair of electrons, partially satisfying all three atoms. The water molecule thus consists of an oxygen atom and two hydrogen atoms, sharing electrons in such a way that the oxygen has a partial claim on 10 electrons arranged in a 2 + 8 form, while each hydrogen has a partial claim on 2 electrons, filling its first energy level.

Consider the *oxygen molecule,* O_2. Two oxygens have got together here and shared *two* pairs of electrons to form a molecule. Each atom has 6 electrons in its outer level, but by sharing 2 of the other atom's electrons, each atom has partial claim on 8 electrons.

STUDENT: But how does this sharing of a pair of electrons keep the atoms together to form molecules?

PROFESSOR: Let's try looking at an analogy:

Frick and Frack Farkel facetiously fling Frisbees fetched from father Farkel, frivolously frittering away fleeting fortnights.

The Farkels are held together by the action of exchang-

Figure 5.30 *Frick and Frack Farkel flinging Frisbees*

ing the Frisbees, just as the atoms are held together in the form of molecules by sharing the pair of electrons (Figure 5.30). If we look at the rest of the atom aside from the shared electrons, it is positively charged. The shared electrons between the positively charged "rest of the atoms" forms the familiar + − + configuration.

STUDENT: OK. I can see that molecules can be formed from atoms by having the *atoms share a pair of electrons* and this may be called a *covalent bond*. What's bothering me is that since atoms are electrically neutral, and no electrons really change hands, molecules must also be electrically neutral. How are we ever going to assemble electrically neutral molecules to form solids, which are supposedly held together by electric forces?

PROFESSOR: Very cleverly! There are two different ways.

A. Natural Dipole

If the atoms that are bonded together covalently to form molecules hold onto their electrons with significantly different strengths, the resulting electron cloud may be quite asymmetric. A good example of this is H_2O. The oxygen atom's nucleus has a much greater positive charge than the hydrogen atom's nucleus, so the shared electrons spend more time near the oxygen atom than near the hydrogen. This leads to an electron cloud as shown in Figure 5.31.

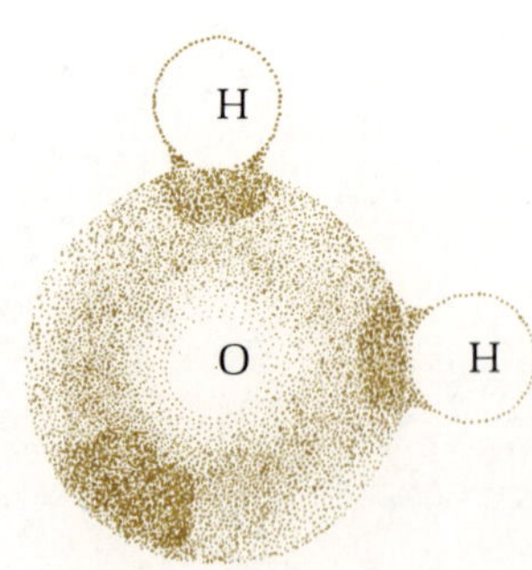

Figure 5.31 *Water molecule electron cloud*

Because the electrons spend so little time on the "other" side of the hydrogen atom, the hydrogen nucleus with its positive charge dominates this area. The oxygen atom, on the other hand, has many electrons around it, so the negative charge dominates this area. Thus, the water molecule is electrically neutral, viewed as a whole, but actually has positive and negative regions. To form a solid, all we have to do is assemble these molecules with positive regions next to negative regions (see Figure 5.32).

Figure 5.32 *Water molecules assembled to form ice*

B. Induced Dipole

Even if the atoms which bond together covalently are identical, which leads to a perfectly symmetrical electron cloud, it is possible to assemble these molecules together to form a solid.

Recall in Chapter 4, we discussed the induced charge phenomenon in which a negatively charged rod repelled negative charges to the far side of a conductor, leaving opposite charges close together, which led to a net attraction (Figure 5.33).

Similarly, if we put two iodine atoms together to form an iodine molecule, as one of atom 1's outer electrons comes around between atom 1 and atom 2, the repulsion between this electron and one of atom 2's electrons causes atom 2's electron to move to the other side of atom 2. Thus, the region to the left of atom 1 is temporarily less negative (or even positive) while the region to the right of atom 2 is temporarily more negative. As the electrons keep moving, this situation is only temporary and doesn't persist for long. Later, the situation may look like that illustrated in Figure 5.34.

Now the atoms have essentially reversed roles, with atom 2's electron repelling atom 1's electron over to the left side of atom 1. Now the left side of the molecule is more negative and the right side less negative.

Figure 5.33 *Induced dipole analogy with induced charge*

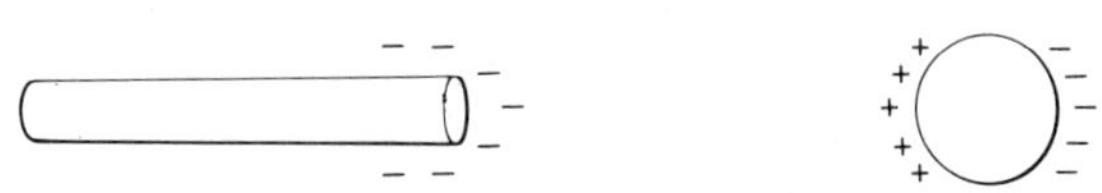

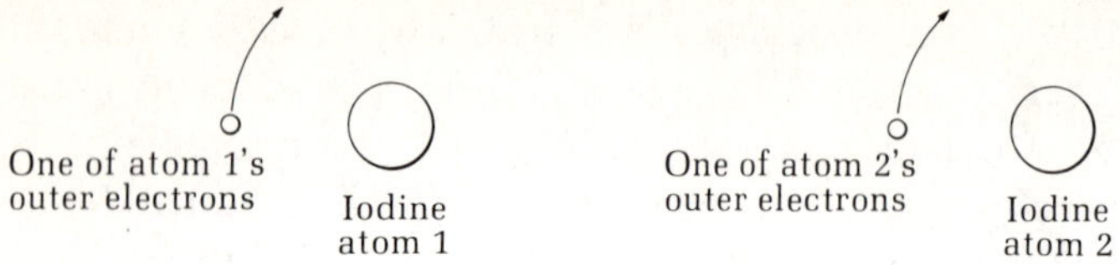

Figure 5.34 *A later look at the I_2 molecule*

And so it goes. *The electrons get "in step" and create instantaneous charge asymmetries,* leading to instantaneously positive and negative regions of the molecule. We can now assemble these very carefully to form a solid.

STUDENT: Wow! I can see the natural dipole fairly well because of the definite positive and negative regions, but the induced dipole seems kind of flimsy. Don't solids formed this way tend to fall apart rather easily?

PROFESSOR: Your intuition is pretty good. Ionic crystals are generally most tightly bonded, with natural dipoles slightly weaker because their charges are not fully transferred, and induced dipoles weakest of all because their charges vary from side to side of the atom. You must recognize that the figures have focused our attention on only one pair of electrons in the outer energy level getting "in step," when actually they all influence each other; so it really isn't quite as bad as it looks.

STUDENT: Now that we've gone over molecules and the way they are held together, what about their potential well structure?

PROFESSOR: It's a little more difficult to envision the potential energy wells associated with molecular crystals because of the more complicated structure of molecules compared to the simplicity of ions, and the less well-defined nature of the bonds, particularly that of the induced dipole, which is constantly changing polarity. Nevertheless, if we were able to "freeze" the action for an instant and figure out the potential and kinetic energies for the molecules in a molecular crystal, we'd get a potential energy diagram like Figure 5.35.

The chief observable difference is that the potential energy wells are not as deep as they were in the ionic crystal, which means that molecular crystals will generally have a lower melting temperature than ionic crystals. There is another difference which we should discover in Experiment 5.3.

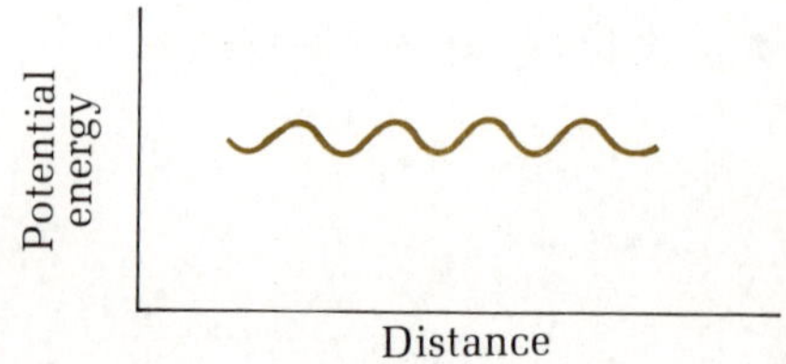

Figure 5.35 *Molecular crystal potential energy diagram*

EXPERIMENT 5.3
MOLECULAR CRYSTAL COOLING CURVE

Recall Experiment 3.2, in which you melted some salol on a glass slide by heating it with a match, and it didn't recrystallize until you dropped a piece of solid crystal into the liquid. We're going to take a closer look at this phenomenon and plot the cooling curve for salol.

The experimental procedure is very similar to that for Experiment 5.1, ionic crystal cooling curve.

1. *Melt the salol completely by simply holding the test tube containing salol under a running hot water tap. (The temperature of hot water may be 60 to 80°C.)*

2. *Insert the thermometer into the molten salol and begin taking temperature readings at 30-second intervals, being careful not to take the thermometer out, or even move it around during the cooling process. Continue recording temperature readings until the salol is completely solid and the temperature has dropped to 30°C or less. As it cools, you should occasionally lightly touch the outside of the test tube to see if it is still increasing the temperature and hence the kinetic energy of its surroundings.*

3. *Make a graph showing the variation of temperature with time as the salol cooled, similar to the one you made for Experiment 5.1 (see Figure 5.36).*

4. *There should be some significant differences between this graph and the one you constructed for Experiment 5.1. Before you read any further, note the differences and try to explain them in terms of the differences in structure between ionic crystals and nonionic crystals. Now try to divide the cooling curve into regions and explain the variation of both kinetic and potential energy in each region, paying particular attention to the conservation of total energy for the salol and its surroundings.*

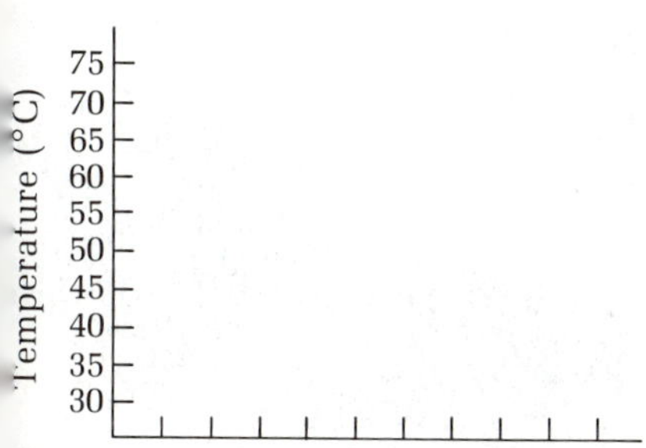

Figure 5.36 *Graph format for Experiment 5.3*

Experiment 5.3 should have yielded a graph like the one illustrated in Figure 5.37.

There are two fundamental differences between this cooling curve and the ionic crystal cooling curve we saw earlier.

1. The plateau temperature is lower, indicating a weaker bond. Although Experiment 5.1 did not involve a real ionic crystal, it was a molecular crystal with a higher

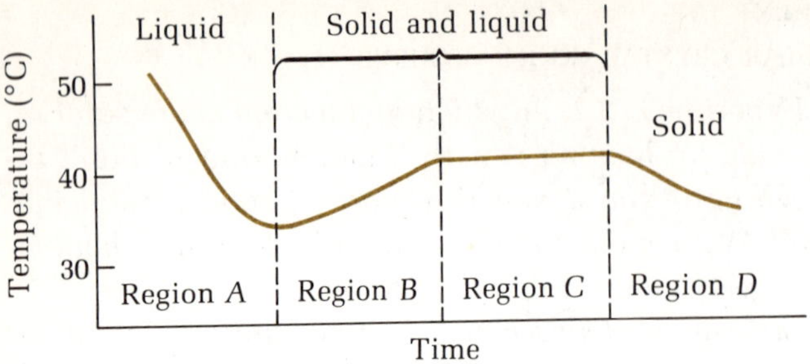

Figure 5.37 *Salol cooling curve*

melting temperature and, therefore, it had deeper potential wells.

2. There is a funny little "dip" between regions *A* and *B*, where the temperature goes below the usual melting temperature. This phenomenon is called *supercooling,* and it was rather perplexing when we did Experiment 3.2 with the salol on the glass slide and it refused to recrystallize. The basic reason it did not solidify when it reached the proper temperature is that salol molecules are extremely complex natural dipole molecules, and they must be oriented in just the proper way before they can get together to form a proper crystal (Figure 5.38).

If too small a fraction of the salol molecules happen to be oriented properly when the crystallization temperature is reached, the molecules simply stay in the liquid form and continue to give off kinetic energy to the surroundings. As the temperature continues to drop, solid formation becomes more difficult, because the molecules move around less as they lose kinetic energy, and are less likely to move into the correct positions.

At the very bottom of the dip, you undoubtedly noticed the formation of some solid. The potential energy released when the molecules go from the far-apart liquid state to the close-together solid state is divided between increasing the kinetic energy of the surroundings and increasing the kinetic energy of the salol itself to bring it up to the proper crystallization temperature. So, in region *B*, PE_{salol} decreases, KE_{salol} increases, and $KE_{surroundings}$ increases.

Figure 5.38 *Salol molecular model†*

†Note: Models portray atoms as hard, static spheres. You must always keep in mind that an atom's outer surface is an electron cloud, and atoms are in constant motion.

Ionic crystals don't have this problem, because ions are quite simple structures and require no particular orientation to get together and form solids. Going back to Experiment 3.2, for what reason did the salol finally form a crystal in the test tube while it didn't on the glass slide? Because there was so much more salol in the tube, the chances of two molecules stumbling onto the right orientation were much greater. In Experiment 3.2, putting in the "seed" crystal established the pattern by fixing some of the molecules and supplying the potential well structure into which the other molecules could then fit themselves. The reason that the salol wouldn't crystallize onto other things put in the melt, even other crystals, is that none of them supplied the right pattern.

SHARP STUDENT: Aha! Not only does this explanation clear up Experiments 3.2 and 5.3, I think it points to an explanation of Experiment 3.1's home-grown crystals.

PROFESSOR: Oh, you've gotten some results on your home-grown crystals?

STUDENT: Yes, and the results were a little disturbing until I thought about them in light of the salol experiment explanation. Here are the results:

POWDERS	CRYSTALS
Blue and white	Blue and clear
Purple and white	Purple

The white powder was the same in both jars, and it has been bugging me that it formed a separate crystal with the blue powder but not with the purple powder. Well, since the salol crystals wouldn't form on other crystals, I figure that the blue and clear crystals simply had such radically different structures that they wouldn't grow together, while the purple and clear crystals were similar enough that they could grow together.

PROFESSOR: Wow! Not only must the structures be different, but so must the potential energy well spacing and depth, since the potential energy structure is a reflection of the physical structure.

5.5

Molecular Noncrystals

We assembled structurally simple ions to form ionic crystals, and more complicated molecules in a repetitive way to form molecular crystals. Next, let's assemble "different" molecules to form an irregular and, hence, non-crystalline solid. How do we go about getting "different" molecules? There are basically two ways:

A. Form Different Sizes and Shapes of Molecules from the Same Atoms

A prime example of this is the sulfur atom, which can form a molecule in the shape of a "puckered ring" of 8 sulfur atoms or various "chains" ranging in length from 2 to 16 atoms. As crystalline sulfur is melted, it turns to a liquid with its molecules in the puckered ring of 8 form (Figure 5.39). As the liquid temperature is increased, the "rings" come apart and form the "chains" of sulfur atoms of various lengths from 2 to 16 atoms.

B. Mix Together Molecules Made from Different Atoms

Paraffin consists of a mixture of long chain hydrocarbons of different lengths. Hydrocarbon chains are the backbone of organic chemistry, and contain a chain of carbon atoms covalently bonded together, with enough hydrogen atoms also covalently bonded to the carbons so that the carbons and hydrocarbons all get a partial claim on enough electrons to fill their outermost energy level.

Since the basic building blocks of molecular noncrystals are not regular, we cannot expect the potential well structure to be regular either. The solid potential energy well structure should look pretty much like the liquid potential energy well structure, except the solid potential energy should be lower (see Figure 5.40). Because the potential

Figure 5.39 *Sulfur "puckered ring" of 8 atoms and chains of 2, 8, and 12 atoms*

energy wells of the liquid and solid are similar, there is no sharp dividing line between the solid and liquid, and thus there should be no sharply defined melting temperature indicating that the depth of the potential energy wells varies in the solid because they are not uniform.

On the basis of Figure 5.40 and the conservation of energy, predict what a noncrystal cooling curve would look like, then we'll do an experiment to test the prediction.

EXPERIMENT 5.4

PARAFFIN COOLING CURVE

This experiment is extremely similar to Experiments 5.1 and 5.3, so the procedure is simply to melt the paraffin and raise its temperature to at least 80°C, reading the temperature at 30-second intervals as it cools to 40°C or below. Construct a temperature-time graph and compare it to your prediction made above. SAFETY precautions and equipment cleaning instructions are the same as for Experiment 5.3.

Since paraffin is a mixture of several different molecules, it does not have a definite melting temperature. It simply softens or hardens, depending on the temperature.

Let's do another experiment to illustrate the other class of noncrystals, formed by different sized molecules made of the same atoms.

EXPERIMENT 5.5

SULFUR'S JEKYLL AND HYDE ACT

This experiment can be smelly and dangerous if you're not careful, so watch how sulfur is a real double threat.

1. *Pour some water into a small beaker; fold a piece of filter paper and put it into a funnel resting in another beaker (see Figure 5.41).*

2. *Slowly and carefully heat the sulfur in the test tube to a temperature just above 200°C. (Keep the flame away from the top of the tube!)*

3. *Pour just a little (several drops) of the molten sulfur into the water-containing beaker, and pour the rest onto the filter paper in the funnel.*

4. *After just a few minutes, the sulfur in the filter paper will begin to solidify by forming a crust over the top of the*

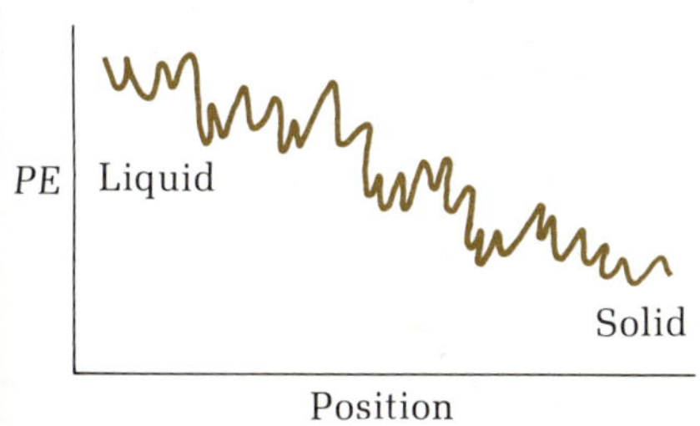

Figure 5.40 *Potential energy diagram for a molecular noncrystal*

cone. When it starts to form this crust, take the filter paper out of the funnel and open it up, exposing the hardening sulfur.

5. *Drain off the water from the other beaker.*

6. *Examine the two forms of sulfur. Try to explain how the cooling process must have influenced the molecules to form the types of solids they did.*

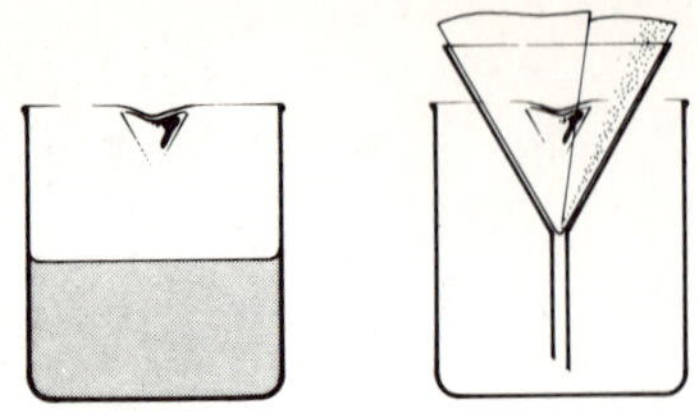

Figure 5.41 *Sulfur cooling apparatus*

5.6

Other Solids

Unfortunately, our discussion of solid matter has had to leave out an extremely important class of solids—metals. They had to be left out because to discuss them to the depth we have discussed the other solids would have required an extensive journey into quantum mechanics. Although it is a very interesting journey for some, it requires more time and mathematical skill than most nonscientists have readily available. Just to skim the surface, metals are said to consist of an array of positive ions in which their outermost electrons are almost completely free to roam throughout the metal, like an "electron gas." This explains the electrical conductivity very nicely, but it leaves us with the question: *How* does the electric force hold metals together to form solids? Although the mathematics may be beyond us, we do know one thing from experience: however the metallic bond works, it surely is strong, because metals don't come apart very easily.

Summary

In an attempt to put together the various kinds of solid matter, we formulated a theoretical model for ionic crystals based on what we knew about crystals, ions, the distance between ions in a crystal, and the electric force holding the ions together. The first model was a static one in which the ions didn't move. An experiment showed this model to be inadequate, because the model couldn't explain the "flat spot" in the cooling curve as an ionic crystal-like substance cooled down from a liquid to a solid form. To get some concepts for a better model, we next investigated energy. We found kinetic energy—energy of motion—and potential energy—energy of position. There are various kinds of potential energy. We looked at gravitational, elastic, and electric potential energies. Using gravitational potential energy and kinetic energy of a pendulum, we showed experimentally that the total energy, kinetic energy plus potential energy, of an isolated system is constant. Next, we constructed a kinetic ionic crystal conceptual model in which the ions are all in motion and therefore have kinetic energy, which is proportional to the temperature. The "flat spot" in the cooling curve occurs when the ions jump from the higher potential energy liquid to the lower potential energy solid. Because the ionic crystal is a regular arrangement of ions, an ionic crystal's potential energy diagram is a regular series of "wells."

Molecular crystals were studied next. Whereas ions completely lose or completely gain electrons, molecules consist of atoms sharing electrons. If the electrons are shared unevenly, there develop slightly positive and slightly negative regions on the molecule, and negative regions may be "stuck" to positive regions by the electric force to form molecular crystals. This is called a natural dipole bond. If the electrons are shared evenly, an induced dipole bond may be formed as the electrons get "in step" to momentarily create positive and negative regions which "stick" together only momentarily. Because molecules can become quite large and complex, some liquids cool below their normal crystallization temperature (supercool), and then their temperature rises when the solid begins to form.

Molecular noncrystals may be formed from different molecules made from the same atoms (e.g., sulfur with molecular "chains" from 2 to 16 atoms long) or from molecules made from different atoms (e.g., paraffin with different hydrocarbons). Because these molecules are not regular like crystal molecules, the potential energy well structure is not regular either and there is little distinction between the liquid and the solid. Thus, you wouldn't expect a well-defined melting temperature, which was verified by experiment.

Another class of solids—metals—cannot be dealt with in any detail, but the outer electrons from metal atoms roam around like an electron "gas" to hold the positive metal ions together.

The conceptual model we have developed for solid matter should allow us to explain many observations in terms of what is going on at the ionic or molecular level. In the discussion questions that follow, *think like a molecule!* Think of yourself as a molecule or ion, and discuss the questions in terms of molecular charge distribution, the degree of complexity of the molecules, the average distance between molecules as a measure of potential energy, the random motion of molecules as a measure of kinetic energy and hence temperature, the depth of the potential wells as a measure of melting temperature, and so forth. Be careful to avoid trying to apply large concepts to small situations. (For example, at a molecular level, friction is *not* rubbing. Our cloud model of the atom says that "rubbing" makes no sense at the atomic level.)

The following questions are meant to be *very* challenging; however, you should be able to come up with the answers by using what you just learned in this chapter. If the answer doesn't occur to you immediately, let it turn over in your mind for a while as you search for some molecular characteristic which would result in the observed behavior of the solids.

1. How might glue work at the molecular level? Note that some glues don't work well at all on plastics.
2. What molecular properties allow a ball to bounce, and why do superballs bounce so well? (What force sends them back up?)
3. Some solids (like table salt) stick together when it is humid; other solids (like metals) stick together in the vacuum of space. Why don't all solids stick together when you shove them together? (What do surfaces look like at the molecular level?)
4. Explain the molecular basis for dirt particles sticking to things. (Hint: Sometimes dust picks up an electric charge by friction as it goes through the air.)
5. What molecular characteristics make Teflon so slippery? (What must be the charge distribution of the molecules?)
6. Vulcanizing rubber involves mixing the rubber with sulfur and heating it up. How could this procedure influence the rubber to remain stable over a wider temperature range? (Remember the sulfur chains.)
7. For what reason does freezing preserve food? (Think about KE and formation of new molecules.)
8. How would melting and recrystallizing an ionic salt tend to purify it? (Think about molecules of "impurities.")
9. Rub your hands together, and you'll notice that they

seem to heat up; their temperature and hence their *KE* has increased. Explain friction in terms of the electric force and molecular structure of solids.

10. Explain cleavage of crystals, particularly of NaCl. (Draw a diagram and shove all ions down one level.)
11. Why do *some* solids (e.g., hair, toast, and paper) never melt, but instead form new molecules when heated. (Think about bonds joining atoms to form molecules, as opposed to bonds joining molecules to form solids.)

Questions

This may be used like a programmed text by referring to the indicated questions while reading the text. (Note: One of the numerical answers given is incorrect; perhaps you can find it.)

5.1

1. Is it possible for Na^+ and Cl^- ions to "overlap," or do they approach only close enough to "touch"?
2. Explain what happens to the ions if they do overlap as a result of the structure of their charges.
3. Explain briefly how °C differs from °F.
4. Describe the static ionic crystal conceptual model.
5. What feature of the ionic crystal cooling curve (Experiment 5.1) cannot be explained easily by the static ionic crystal conceptual model?

5.2

6. Construct examples based on your experience to show that energy increases with a body's speed, mass, and height.
7. Define kinetic energy.
8. Define gravitational potential energy.
9. Explain the conservation of energy.
10. What are the units of energy?
11. What is the kinetic energy of a 0.01 kg snail, crawling along at 2×10^{-4} meters/sec? (2×10^{-10} joules)
12. What is the kinetic energy of a 4000 kg space ship, zipping along at 8000 m/sec? (1.28×10^{11} joules)
13. What is the kinetic energy of a 60 kg water-skier going 10 m/sec? (6000 joules)
14. What is the potential energy of a 70 kg window-washer with respect to the ground 150 meters ($\approx$ 50 stories)

below? (Use $g = 10 \text{ m/sec}^2$ in all *PE* problems.) (1.05×10^5 joules)

15. What is the potential energy of a 2×10^{-3} kg insect with respect to a table 2 meters below? (4×10^{-2} joules)
16. What is the potential energy of a 4000 kg helicopter with respect to the ground 3000 meters below? (1.2×10^8 joules)
17. What is the total energy of a 5 kg road-runner running along a plateau 10 meters above the flat at a speed of 8 m/sec? (660 joules)
18. What is the total energy of a 2×10^{-3} kg insect flying 2 meters above the ground at a speed of 1.5 m/sec? (4.225×10^{-2} joules)
19. Explain the idea of a potential energy well.
20. How is elastic potential energy related to gravitational potential energy?

5.3

21. Explain the concept of electrical potential energy.
22. Explain how electrical potential wells fit into the second model of ionic crystals.
23. How could a thermometer measure the average kinetic energy of a collection of molecules?
24. Draw and explain the energy diagram for an ionic crystal solid and liquid.
25. Explain the energy variation in each region of the cooling curve obtained in Experiment 5.1.

5.4

26. Explain the idea of a covalent bond.
27. What are molecules?
28. Explain the two ways in which molecules are hooked together to form molecular crystals.
29. What are the differences between the potential energy diagrams for molecular crystals and ionic crystals?
30. Explain what is happening to all the different kinds of energy in the supercooling region of the molecular crystal's cooling curve.
31. What is the structural reason for this supercooling phenomenon?
32. Explain the results of Experiment 3.1, home-grown crystals, in terms of the structure and potential energy diagrams of molecular crystals.

5.5

33. List two ways that noncrystals could be formed from molecules.
34. Draw and explain the potential energy diagram for a noncrystal.

5.6

35. Why don't we study metals in more detail?

Quotes to Ponder

1. "Ford Motor Company has announced that its 1973 model light trucks now on sale have wheels painted with a new process using a dry powder instead of liquid paint.

The coatings, from a new family of acrylic-based powder paints, are applied with an electrostatic process which essentially magnetizes the wheels to attract the paint.

The coating is then baked to a hard, glossy smooth surface. . . ." (Oakland Press, October, 1972)

CHAPTER SIX

SOLID MATTER + ENERGY = SOLID JUNK

Now that we know some things about solid matter, let's take a look at our mostly solid planet Earth. Where did this big hunk of solid matter come from? How is it arranged at this time? We seldom seem content with the current arrangement of the solids on the earth, so we are constantly shifting them around (like some people with furniture). This rearrangement requires energy. Where does this energy come from? Finally, when we get tired of some particular solid objects—better known as "junk"—we throw them "away"; but what really happens to them?

To get answers to these questions, we'll have to dip into various branches of physical science: astronomy, cosmology, geology, and chemistry. Furthermore, we'll have to make use of a number of things we learned earlier about solids and how they fit together.

6.1

Where Did All the Solid Matter Come From? Cosmology's "Big Bang"

The question of the origin of the universe is one that has always fascinated people. The branch of science that deals with theories of the origin and development of the universe is called *cosmology;* the branch of science that provides the experimental data which these theories must explain is called *astronomy.*

There has never been any shortage of theories about the origin of the universe. The major difficulty has been obtaining astronomical data of the right kind to either confirm or deny the theories. The universe is not a very cooperative laboratory. It is not possible to hold all but one of the variables constant, run experiments, and take data the way we have done with our laboratory experiments in earlier chapters. Instead, astronomers must be content with intercepting and observing the electromagnetic radiation (e.g., visible light, x-rays, infrared, ultraviolet, and radio waves) which is given off or reflected by asteroids, planets, stars, comets, galaxies, and so forth.

Historically, the fate which befalls most theories of the origin of the universe follows a sequence like this:

1. A theory is constructed which explains all the known astronomical phenomena to within the limits of accuracy of present measurements.
2. Astronomers either improve the accuracy of their instruments; or they discover new phenomena.
3. The theory cannot explain the more accurate data or the new phenomena.
4. The theorists go back to the drawing boards for a new theory.

Furthermore, the big joker in astronomy is time. Man has

been making detailed, systematic astronomical observations for, at most, a few thousand years. Yet the age of the universe is estimated to be about *10^{10} years!* We are in the unfortunate position of observing the equivalent of about a minute in the life of a 20-year old and trying to figure out what his life history was and what his future will be.

Of all the cosmological theories, the one currently thought to be the best is the "Big Bang" theory, finally formulated in 1948 by George Gamow. This theory says that about 10^{10} years ago, all the matter and energy in the universe were concentrated into a single, hot (about 10^{16}° C), dense (about 10^{25} grams/cm^3) fireball (Figure 6.1).

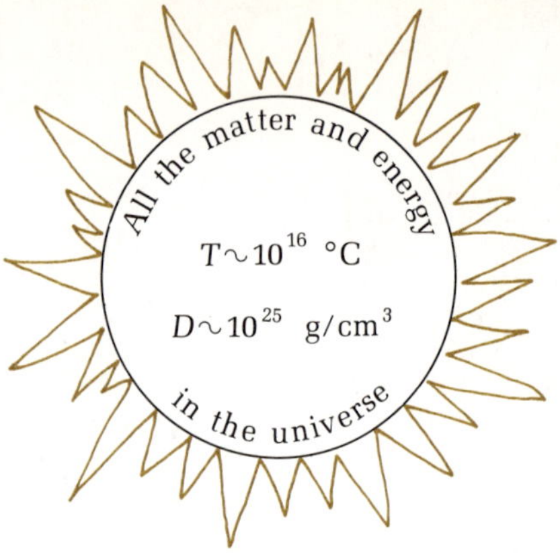

Figure 6.1 *Primeval fireball*

Although it is difficult to imagine anything as hot and dense as this fireball, we might try constructing some analogies. For example, the temperature of this fireball was 100 million times that of our sun, and a pitcher full of the fireball material would contain more mass than the earth.

According to the theory, this hot, dense material was in the form of protons, neutrons, and electrons which did *not* combine to form nuclei or atoms because of their high kinetic energy and density. There were many, many particles in a small space with tremendous amounts of kinetic energy and hence activity.

Although the reason is not at all clear, the primeval fireball *exploded,* scattering protons, neutrons, and electrons in all directions. In the initial stages of the explosion, some protons and neutrons got together to form helium nuclei (2 protons + 2 neutrons), but the expansion carried things apart too fast to form any more massive nuclei than helium, and it is theorized that the amount of helium formed was only about 10% of the amount of hydrogen.

How about the energy of the expanding fireball? Recall gravitational potential energy and the way it depends on position. The "higher" any object is, the greater its gravitational potential energy. The farther away from the earth, the greater its potential energy. In other words, since all masses attract each other gravitationally, you must work against this attractive force to pull any objects farther apart. Since this fireball did expand, *gravitational potential energy did increase.* We believe that conservation of total energy was working in those times also, so there had to be some kind of energy decrease to keep the total energy constant. The *kinetic energy and, therefore, the temperature of the fireball decreased* as it expanded. As the kinetic energy decreased

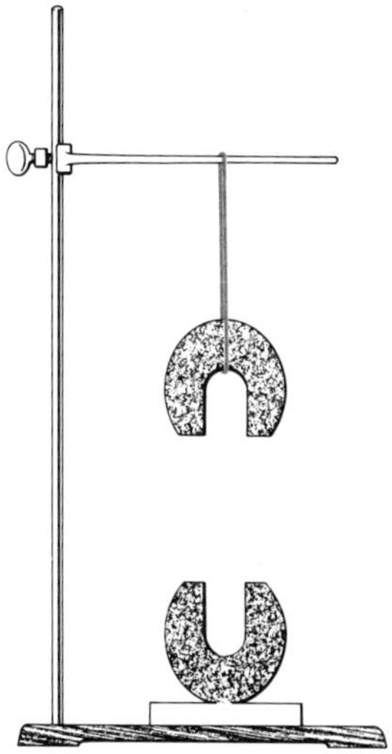

Figure 6.2 *Magnet suspended above another magnet by a rubber band*

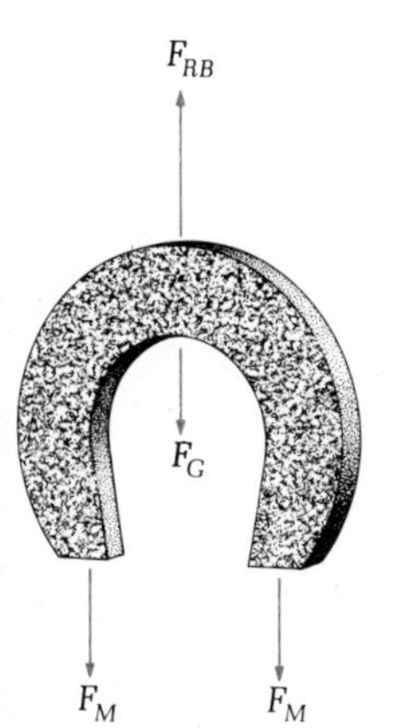

Figure 6.3 *Forces on a suspended magnet*

and things slowed down a bit, some hydrogen nuclei and helium nuclei captured electrons and became hydrogen and helium atoms. The earth contains many more atoms than just hydrogen and helium, so we'll have to follow the Big Bang theory a bit further to see where the more massive atoms are formed.

If you've ever seen an explosion or the aftermath of an explosion, you're well aware that explosions do not scatter material in a completely orderly, uniform way. It is believed that the expanding fireball contained nonuniformities, or "lumps," in which quantities of hydrogen and helium atoms were concentrated. In a beautiful but intricate way, some of these "lumps" developed into stars. This development is rather complex, so let's begin by looking at a simple analogy of the whole process. Imagine a magnet suspended above another magnet by a rubber band, as shown in Figure 6.2.

Presuming that the north poles of the magnets were lined up with the south poles, they would stick together if it were not for the rubber band holding the top magnet above the bottom one. How could you get the magnets together? You could simply raise the top magnet a bit and let it go. The earth would attract it gravitationally, pull it down, and stretch the rubber band; and if it stretched enough, (click!) the magnets would come together. From an energy standpoint, you would trade gravitational potential energy for enough kinetic energy to overcome the elastic barrier. A diagram of the forces on the top magnet is shown in Figure 6.3.

F_G is the gravitational force exerted on the magnet by the huge mass of the earth. F_{RB} is the force of the rubber band, pulling the magnet back up. F_M is the magnetic force, trying to pull the magnet down. The magnetic force is not very strong when the magnets are far apart, but it becomes quite strong when they are close together, stronger than the restraining force of the rubber band trying to keep the magnets apart.

Now let's see how this analogy applies to the formation of stars. Rather than have two magnets in the vicinity of the earth, let's look at two hydrogen atoms in the vicinity of a large number of hydrogen and helium atoms as shown in Figure 6.4.

The "outer" hydrogen atom is gravitationally attracted to the large mass of atoms just as the upper magnet is gravitationally attracted to the earth. Before the "outer" hydrogen

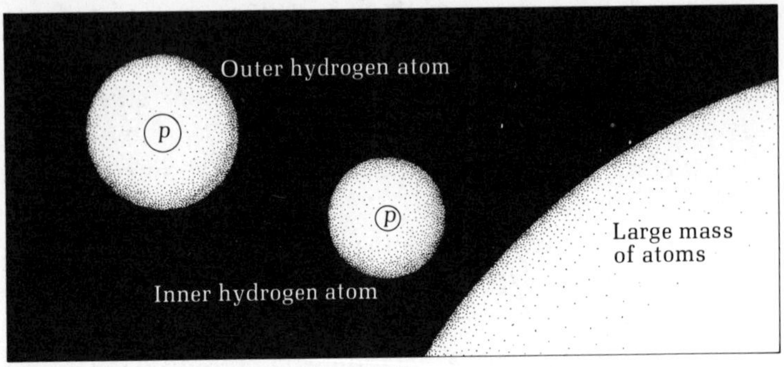

Figure 6.4 *Two hydrogen atoms near a large mass of atoms*

atom bangs into the "inner" hydrogen atom, the electrons will repel each other out of the way; and the protons will face each other and build up a repulsive force, just as the rubber band tried to keep the upper magnet from hitting the lower magnet. Just as the magnets will stick together if they get close enough, so will the protons stick together if they get close enough. Just as the magnetic force holds the magnets together, the protons are held together by the nuclear force. (Recall that the nuclear force is very strong, but very short range.) A force diagram for the proton in the "outer" hydrogen atom is shown in Figure 6.5.

F_G is the gravitational force caused by all the other atoms, F_E is the repulsive electric force between the proton and the other proton, and F_N is the short-range nuclear attractive force generated by the other proton. Will the protons get close enough for the nuclear force to overcome the electric force so that the protons will "stick" together? That is the same question we asked about the magnets. Again, the answer lies in the energy. If there is enough gravitational potential energy that can be turned into kinetic energy to overcome the electric force "barrier," then the protons will stick together. The chief factor influencing the gravitational potential energy is mass. If the bunch of atoms in the "lump" contain enough mass, the proton will quickly move in, with sufficient kinetic energy or temperature to get close enough to "stick" to another proton. This "sticking" process is called *fusion* and it is going on in stars continuously.

The fusion process which occurs first in the development of a star is called the *proton-proton chain*. Four protons are stuck together to form a helium nucleus with lots of kinetic energy, plus some electromagnetic radiation, plus some other particles. This process is illustrated in Figure 6.6.

(The "other" particles are two positrons, which are just like electrons except they have a positive charge, and two

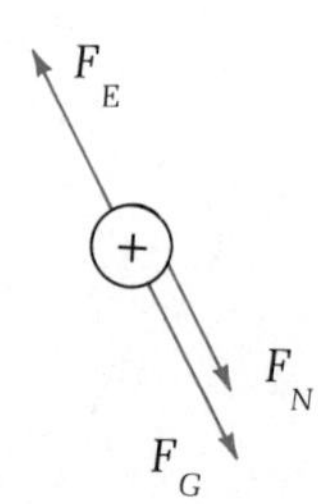

Figure 6.5 *Forces acting on the proton in a hydrogen atom*

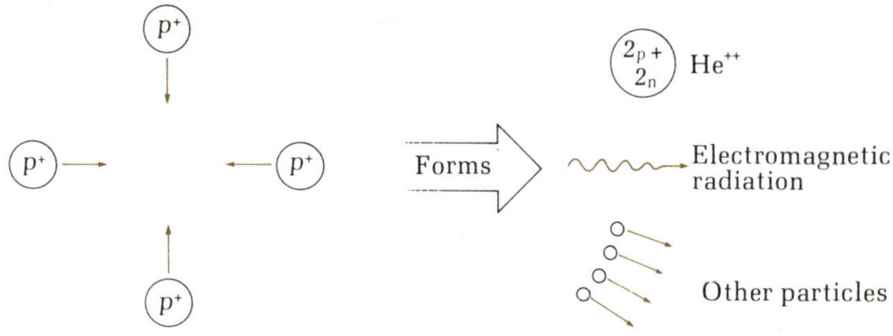

Figure 6.6 *The proton-proton chain*

electron neutrinos, strange particles which are not fully understood. Although high-energy physicists find these other particles interesting, they are not necessary for our purpose here, so they will not be mentioned further.)

Once these highly active, energetic helium nuclei begin forming, other more complicated fusion processes occur between the helium nuclei and protons, neutrons, and other helium nuclei. These more complicated fusion processes produce heavier nuclei and give off electromagnetic radiation and other particles just like the proton-proton chain. Because these heavier nuclei have greater amounts of positive charge, their electric barrier is more difficult to overcome; so more kinetic energy is needed, and thus the temperature in the star's interior must increase before these heavier nuclei may be formed. Thus, a star's normal development would start with the proton-proton chain; and its temperature would rise because of the energy of this reaction, allowing heavier elements to be put together. There doesn't appear to be any limit to the size of the nucleus that could be formed, so why can't a star simply continue making larger and larger nuclei? After all, stars do last a long time.

It turns out that a star's lifetime and the size of the nuclei that it forms are closely related; and from this relationship, we'll finally get to the point of where the earth got all its elements. The size of a nucleus that a star may form is limited by the short-range nature of the nuclear force. Let's take a look at some of the protons in a large nucleus.

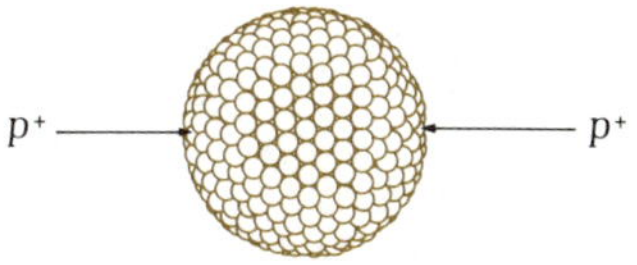

Figure 6.7 *Two protons in a large nucleus*

The two protons in the nucleus shown in Figure 6.7 are relatively far apart, so the nuclear force between them is not as large as the electric repulsion. Thus, although the neighboring protons and neutrons exert enough nuclear force to bind either of these protons, they cannot be bound as tightly as they would be if the protons on the other side were not repelling. Thus, larger and larger nuclei become more and more loosely bound until finally no more protons may be stuck in, because the repulsive electric forces are greater than the attractive nuclear forces exerted by the neighboring protons and neutrons. Because of this size limitation, elements larger than uranium, $^{238}_{92}U$, do not occur naturally. More massive nuclei have been made artificially, but they come apart very rapidly.

There is another consequence of the limitation on nuclear size due to the short-range nature of the nuclear force. As stars continue forming more and more massive nuclei, there comes a point at which the fusion process no longer gives off energy, but instead *requires* energy to form more loosely bound, massive nuclei. This point is reached at the formation of iron, $^{56}_{26}Fe$.

As the star forms elements less massive than iron, energy is given off in the form of electromagnetic radiation; as the star forms elements more massive than iron, energy is *required*. Where does this energy come from? It comes from the smaller nuclei being formed in other parts of the star.

The energy being given off in the smaller fusion processes is no longer all sent out in the form of electromagnetic radiation; some of this energy is now used to form the more massive nuclei. This point is crucial to the stability of the star. When only light nuclei are formed, the electromagnetic radiation given off in the process does two things: First, it escapes in the form of x-rays, gamma rays, light, etc.; second, the electromagnetic radiation bangs into the protons around the outer edges of the star, keeping them from being pulled into the star's center too fast by gravitational attraction. (See Figure 6.8.)

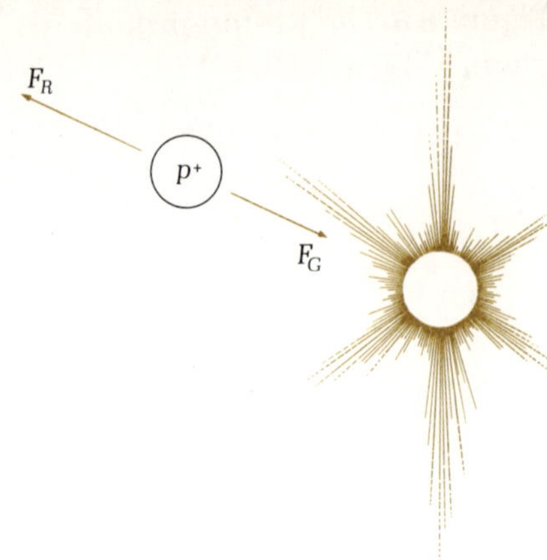

Figure 6.8 *Proton being hit by electromagnetic radiation*

The electromagnetic radiation causes the protons outside the core of the star to remain outside, rather than being pulled in too quickly by gravitation. If the fusion reactions in the star's core slow down, the electromagnetic radiation is reduced, and more protons come in to keep the fusion process going. The electromagnetic radiation acts like a buffer, keeping the star's fuel (protons) out of the way until they are needed.

But now watch what happens as the star develops, gets hotter, and begins to form nuclei more massive than iron. The radiation that formerly went out, keeping the protons out of the action, is now partly being used to make the heavier nuclei in the star's core. The protons are no longer held out as they were, and they eventually rush in, causing the fusion process to proceed much more rapidly than before until finally . . .

the star *explodes*. If the star is not too massive (that is, less than about 1.4 times the mass of our sun) the explosion is termed a *nova*. If the star is more massive than 1.4 times the mass of our sun, it attracts the protons inwardly in an extremely strong way, and a particularly catastrophic explosion, called a *supernova,* takes place.

A supernova scatters the material from the star's core in a way very similar to the way material was scattered from the original Big Bang. The important difference is that the material being scattered by the big bang was hydrogen and helium, while the supernova scatters all the nuclei that the star has been so busily making.

(By the way, the time necessary for a star to develop from first fusion to nova or supernova is believed to be measured in millions or billions of years. Our own sun, for example, is thought to be about 4.5×10^9 years old, and it is estimated that it has another 4 or 5×10^9 years before it will explode in a nova. We still have a little time to get ourselves together.)

Now, just as the material from the Big Bang exploded nonuniformly and cooled down as it did, let's follow what happens to a cloud of material blown out of a star which exploded in a supernova. The cloud begins to come together because of gravitational attraction, just as the original star did. If there is enough mass in the cloud, the nuclei will come together with enough kinetic energy to overcome the electrostatic repulsion, and so fusion will occur and another star will be born. If the cloud comes together with insufficient mass and thus too little kinetic energy to begin the fusion process, a relatively cool body will be formed containing all the elements blown out of the star in the supernova. This is what we would call a *planet* and, specifically, this is one theory of the origin of our planet Earth.

Now that we have used the Big Bang theory and extensions of it to arrive at the formation of the earth, it seems that we should examine the evidence for this theory a little before we proceed to analyze the earth in more detail. After all, theories of the universe have come and gone. What makes this particular theory any better than the others?

The Big Bang theory explains several observations very nicely:

1. *The universe seems to be expanding.* Astronomers have measured the velocities of galaxies in the universe and found them to be moving away from each other, some at nearly the speed of light. The Big Bang theory says that this moving away from each other is a natural result of that initial explosion.
2. *There appears to be about 10% as much helium as hydrogen in the universe.* This finding confirms the theoretical prediction that during the initial phase of the Big Bang, the fundamental particles were cool enough and close enough to form only 10% as much helium as hydrogen.
3. *The background radiation for all directions in space appears to be that of a body whose temperature is −270.3°C.* The Big Bang theory says that the "fireball"—of which we are still a part—should be cooled down to about −270°C by now.

Is this theory the last word on the origin of the universe? Probably not, but just like the other theories we have studied, it is the one which best fits the experimental data available right now. Actually, there is a very interesting modification to the theory which says that the universe will stop exploding and turn around and come together to form the fireball and explode in another big bang all over again. This is the *"oscillating universe"* theory and it depends quite strongly on the total mass of the universe. If the total mass is large enough, the universe will oscillate between expanding and contracting. If the total mass is too small, it will simply continue expanding. It would seem logical, although extremely time-consuming, to add up the estimated masses of all the known galaxies and find out if the universe will expand or oscillate. Unfortunately, it's not that easy because there may be masses "out there" of which we aren't aware.

The way astronomers normally sense the presence of objects in space depends on the objects' ability to radiate or reflect electromagnetic waves, like light. For many years astronomers have been searching the skies with their telescopes, carefully cataloging objects which they can see. We realize that the electromagnetic spectrum contains many waves besides visible light, but only recently have astronomers been able to devise equipment to search for electromagnetic waves from the nonvisible part of the spectrum. A variety of problems, not the least of which is the uncooperative nature of our atmosphere, have prevented infrared, ultraviolet, radio, radar, x-ray, gamma ray, and so forth, astronomies from really flourishing until recently. Even more fundamental problems, however, prevent an accurate estimate of the total mass of the universe. Suppose there were things in the universe which had mass, but didn't emit or reflect electromagnetic waves to a large enough extent to be detected. Bare protons emit no electromagnetic waves, and it has been theorized that another class of objects exist which also emit no radiation. These objects are called *"black holes"* and are thought to be undergoing such a violent gravitational collapse that no radiation can escape. "Black holes" are thus impossible to detect by radiation sensors. Astronomers are rather clever, however; so don't be surprised to read about an experimental proof or disproof of "black holes" and a better understanding of whether we live in an expanding or oscillating universe.

EXPERIMENT 6.1
HEAVENLY BODIES

Get a telescope, a star chart, and wait for a clear evening. Observe and identify some stars and planets and simply ponder such questions as: How big are they? How far away are they? How hot must they be? How many are "out there?" How many other stars must have planets? What is the mass of the earth in comparison to the masses of the stars? How can we think we're such big deals in the face of the overwhelming numbers of heavenly bodies and the vastness of space? Don't get too carried away, but realize that all this has been intriguing astronomers for years.

6.2

How Did the Earth Get Into the Shape It's In? Very Geologically

STUDENT: If we can believe the Big Bang theory, the earth was formed from the debris of stars that underwent a gravitational collapse, called a supernova.

PROFESSOR: Yes, the debris of heavy elements attracted each other gravitationally and came together to form the earth.

STUDENT: The earth must have had enough energy to get fairly hot. What happened to this energy? Was it just radiated away?

PROFESSOR: A great deal of this gravitational potential energy was turned into random motion of the atoms and radiated away, but some of it got the atoms moving enough to encourage the formation of molecules, rather than leaving the atoms in an uncombined state.

SHARP STUDENT: Oh, I see the answer to something I was just going to ask about. I was going to ask how the earth managed to retain some of the lighter elements if its gravity is all that weak. The answer is that many of the lighter elements combined chemically to form more massive molecules, and hence were more liable to stick around.

Let's get down to one key question: How is the material in the earth arranged?

PROFESSOR: OK. Let's start with something we can see and analyze, and try to figure out something about what we can't see directly.

EXPERIMENT 6.2

DENSITY OF EARTH AND ITS CONSTITUENTS

We all know that a great deal of the earth must be rock. Get three different rocks, and let's measure their density, just as we did in Experiment 1.1. Recall:

$$\text{Density} = \frac{\text{mass}}{\text{volume}}.$$

You can measure the mass by using a balance, and the volume by water immersion. You should get densities of 2 to 3 grams/cm³.

What about the density of the earth as a whole? To get the volume, we can find the radius by some geometric means involving the sun and shadows. (We'd get 6.4 × 10⁶ meters if we did.) Assume the earth is a sphere, and calculate its volume based on the formula:

$$V_{sphere} = \frac{4}{3}\pi r^3.$$

(We'd get 1.1 × 10²¹ m³.) To get the earth's mass we'll need this big *balance and a bunch of masses. Oops! Let's try another approach. Since weight is just the gravitational force on an object due to the mutual attraction between Mother Earth and the object, let's equate these two forces and see if we can solve for the mass of the earth.*

$$\text{Weight} = \text{gravitational force},$$

$$w = F_G,$$

$$mg = G\frac{mm_E}{r^2},$$

or

$$m_E = \frac{gr^2}{G}$$

where

r = *radius of the earth* $\approx 6.4 \times 10^6$ m,
g = *acceleration due to gravity* ≈ 10 m/sec^2,
G = *gravitational force constant* $\approx 7 \times 10^{-11}$ nt · m^2/kg^2.

Carrying out this calculation yields a mass of 6 × 10²⁴ kg for the earth. The density of the earth is therefore

$$D = \frac{m}{V} = \frac{6 \times 10^{24}\ \text{kg}}{1.1 \times 10^{21}\ \text{m}^3} \cong 5.4\ \text{grams/cm}^3.$$

STOP! The density of the rocks you collected was more like 3 grams/cm³. Before you go any further, think about what this difference in density implies about the earth's structure.

If you figured that there must be some denser solid below the earth's surface, you're on the right track. In order to get to an average density of over 5 grams/cm³, there must be some pretty dense rocks down below to make up for the light ones we found near the surface.

STUDENT: One thing I have learned about physical science is that speculations like the one above are of very little value without some experimental data. When do we start drilling a hole to the center of the earth to find out what's down there?

PROFESSOR: You may think that's funny, but there have been some serious proposals to drill some pretty deep holes, and some day someone will undoubtedly do it. Fortunately, we have been able to find out a few things about the earth's interior by slightly less direct means. Recall we couldn't see inside a crystal, so we shot x-rays through crystals to see what their structure was. Let's shoot some waves through the earth. Disturbances to the earth, whether on the surface or below, create waves. (Surely you've heard the old joke, the punch line to which is "Don't make waves!") The branch of geophysics which studies earthquake waves and the earth's response to them is called *seismology*. Three distinctly different types of *seismic waves* are recognized:

1. *P-Waves:* Longitudinal waves (in which the medium moves in the same direction as the wave) which travel through solids at a speed of 6×10^3 m/sec and through liquids a bit more slowly.
2. *S-Waves:* Transverse waves (in which the medium moves perpendicular to the wave) which travel through solids at a speed of 3 to 4×10^3 m/sec and not at all through liquids.
3. *L-Waves:* Travel along the earth's surface slower than 3×10^3 m/sec.

Let's put ourselves in the position of geophysicists and see what we can figure out about the structure of the earth's interior from studying some seismic records.

Suppose an earthquake begins at the point marked earthquake source at 1:00 p.m. (see Figure 6.9). The observer located at the other side of the earth detects the arrival of a P-wave at 1:53:20 and detects *no* S-wave at all. On the basis of the wave properties given above, what conclusions can you draw about the material below the earth's surface? *Figure it out before you read on.*

The conclusion you should draw from this hypothetical

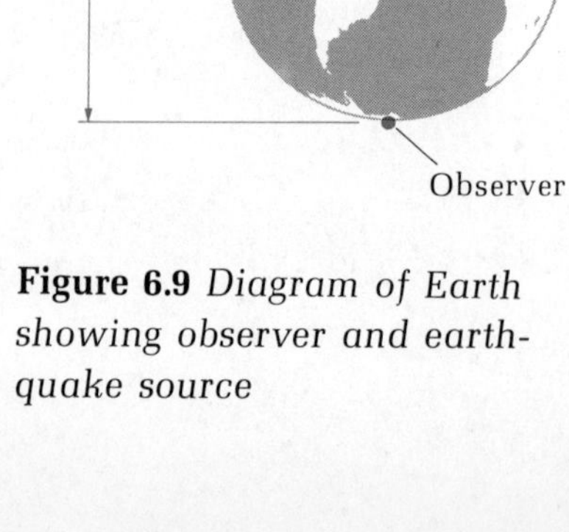

Figure 6.9 *Diagram of Earth showing observer and earthquake source*

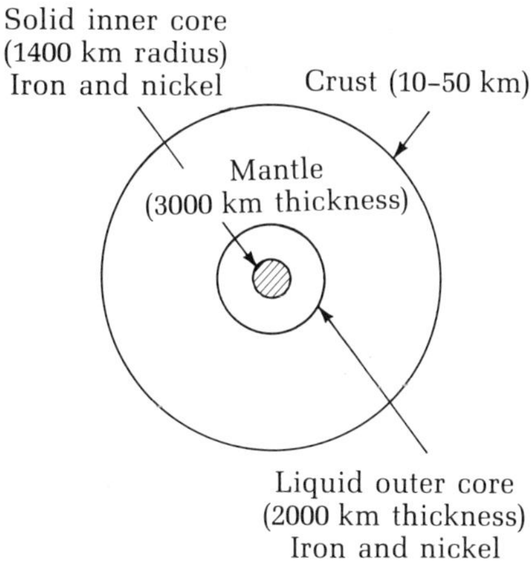

Figure 6.10 *Cross-sectional view of the Earth*

experiment is that there must be some liquid below the earth's surface. First of all, the P-wave was slowed down by something. If the earth were a uniform solid, it would have taken the P-wave only 33 minutes to make the trip rather than almost an hour. Liquid would slow the P-wave down. Secondly, the absence of the S-wave also indicates the presence of a liquid, because S-waves would be stopped by a liquid. Detailed seismological studies (which are very difficult to interpret because of nonuniformities, reflections, etc.) indicate that the earth consists of a solid inner core with a radius of 1400 km, a liquid outer core of about 2000 km thickness, a mantle of 3000 km thickness, and a crust on top of about 10 to 50 km thickness (Figure 6.10).

Both the *outer and inner cores* are thought to consist largely of iron, with perhaps a small amount of nickel. This takes care of the density problem we noticed in Experiment 6.2, because the density of iron is almost 8 g/cm^3 at normal temperatures and pressures, and may go as high as 11 or 12 g/cm^3 in the high pressures of the core.

The *mantle* is thought to consist of fairly dense rocks which are actually above their normal melting temperatures but are a noncrystalline solid because of the pressure being exerted on them.

The *crust* seems to consist of large "plates" of fairly light rock which "float" on the mantle.

As usual, the real test of this hypothetical model of the earth is its ability to explain observations. We saw that the density and seismological observations were central to the formation of this theory, but let's see if we can explain the phenomena of *continental drift* and *earthquakes*. It's not hard to envision the continents on plates, floating away from each other, or to envision two or more plates banging together or coming apart, causing earthquakes. The difficult part is envisioning a *cause* for the motion of these plates. Here's where our background in solid matter comes in. Recall that the mantle is a noncrystalline solid. Noncrystalline solids, like glass and paraffin, are just like liquids, only "stiffer." So there could be "currents" in the mantle, if we were willing to wait long enough for the thick stuff to flow. If time were no object, currents in the mantle could be generated by temperature differences, the earth's rotation, or other causes.

If the current happened to be directed upward at a thin area in the crust, it could actually intrude and shove the crust sideways (see Figure 6.11). The thin spots (10 km or

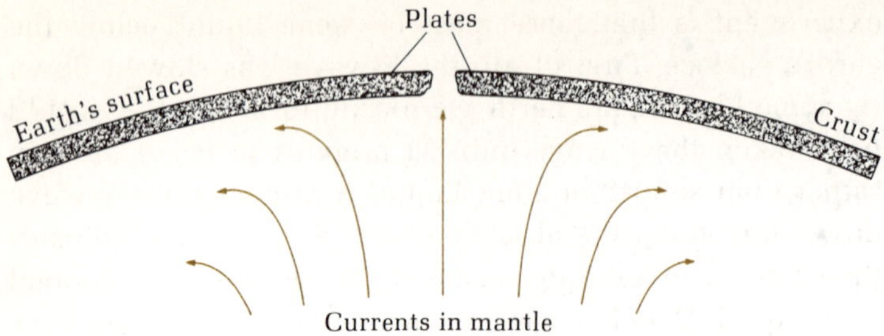

Figure 6.11 *Mantle current shoving plates in crust*

so) in the crust occur under the oceans; so if this theory is any good, we should find areas under the oceans where mantle materials have oozed up to the surface. Such areas have been found, tending to validate this theory.

(There are a number of other topics in geology which you might find interesting, such as mountain building, rocks and minerals, glacier formation, and volcanic eruptions. Please consult the references in the Study Guide for some suggestions of where to look.)

In short, the formation of the earth seems like what would happen to a chunk of iron ore if you pitched it into a blast furnace. The heavy, pure iron would sink to the bottom, the lighter rock would float on top of the iron, and the lightest materials would float like scum on the top. If you think about it, we are living on the scummy surface of a cooled-down planet that was formed from the debris of a supernova, orbiting a medium-sized, middle-aged star in an average galaxy. There is nothing too spectacular about us, but here we are; so let's enjoy it while we can.

6.3

Rearrangement of These Solids Costs Energy, Mostly from the Sun, but Nuclear Power May Be Just Around the Corner

Now that we have some idea of the gross arrangement of the matter that surrounds us here on earth, it seems appropriate to discuss some of the rearrangements of this matter. By the way, one's policy toward rearrangement of the environment is largely influenced by one's culture. There have been cultures whose ethical systems seemed to be dominated by the principle of minimal disturbance of the environment. Western man is at the opposite end of this spectrum, seemingly believing that all things were put on earth to be used by him. The earth is said to be teeming with "natural resources," begging to be used, transformed, rearranged, or whatever by him.

The rearrangements of solids include such activities as mining, earthmoving, construction, manufacturing of hard goods, and transportation. All these activities involve moving some object to some speed (kinetic energy) against friction (heat energy) or to some height (potential energy).

Where does the energy come from to do this work? Let's analyze a few cases.

1. *Mining:* Let's say the ore is chipped out of the ground by men and carried to the surface by a gasoline-powered lift. The energy sources are man and the gasoline.
2. *Earth-moving:* Most large-scale earth-moving is done by machines run by diesel engines. The diesel fuel is the energy source.
3. *Construction:* Smaller construction jobs use man as a primary energy source.
4. *Manufacturing:* Many industrial machines run on electric energy.
5. *Transportation:* Gasoline and other petroleum products are the chief energy sources.

To track down the ultimate source of the energy, we must further investigate the source of the energy used by man, by petroleum products, and by electric energy generators.

Man obtains his energy from the food he eats by using oxygen to convert large, loosely bonded molecules like sugar into small, tightly bonded molecules like CO_2 and H_2O. As we have seen in our study of solids, there is more electrical potential energy in large molecules whose atoms are far apart than in small molecules whose atoms are close together. The decrease in potential energy resulting from the breakup of these large molecules shows up as an increase in the body's internal energy, enabling the body to move its muscles, speak, and think.

The next question is: How does the food happen to have the large molecules, rich in potential energy? It depends on the food, of course, but most natural foods can be traced back to plants. Plants take in small molecules like CO_2 and H_2O and build large, loosely bonded molecules like sugar.

Where do plants get the energy they store in these molecules? From the *sun*. Recall that one of the by-products of the fusion reaction that takes place in stars is electromagnetic radiation. Plants use electromagnetic radiation from the sun to form large molecules from small ones in a process called *photosynthesis*. Thus, the sun is man's ultimate energy source, because energy radiated away by the sun is used by plants to build large molecules, rich in potential energy. Man converts the large molecules to small ones and uses the stored potential energy to carry out various functions.

A grossly oversimplified view of the energy relationship between man and plants and the molecules involved is shown in Figure 6.12.

Gasoline, diesel fuel, kerosene, natural gas, and coal are intermediate energy sources whose ultimate energy source is also the sun. All petroleum products are simply plants which have rearranged their molecules somewhat by being compressed and stored for years, but their molecules are still essentially large and loosely held together. When these large molecules are combined with oxygen to form smaller CO_2 and H_2O molecules, they give up their potential energy. This energy is used to move cars, trucks, airplanes, and boats, heat homes, generate electric energy, and much more. Thus the sun is the ultimate energy source for petroleum products.

Aside from some ways of generating electric energy,

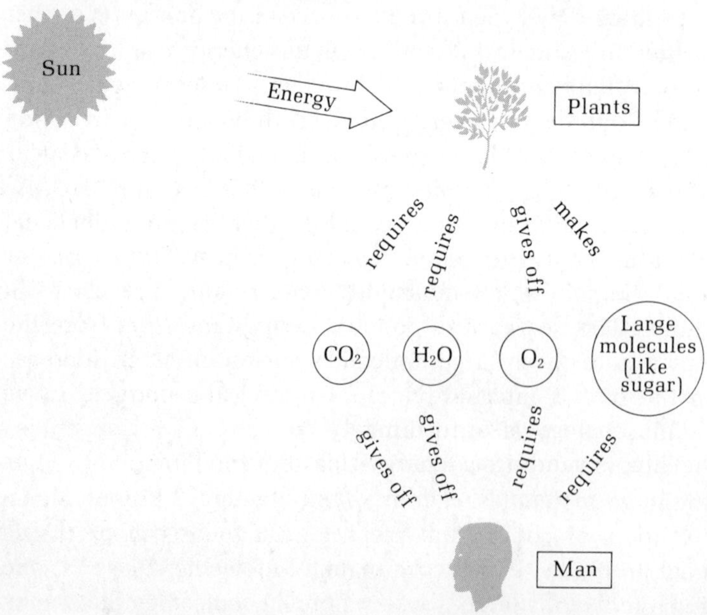

Figure 6.12 *Energy flow from the sun, through plants to man*

Figure 6.13 *Rich uncle T. R. Ump giving money to poor nephew C. H. Ump*

which we will discuss shortly, the sun seems to be the ultimate energy source for almost everything we do. To explore this relationship between the energy given off by the sun and the energy we use, let's look at an analogy involving some members of the Ump family as shown in Figure 6.13.

Rich uncle T. R. Ump makes a great deal of money and sends a small portion of it to nephew C. H. Ump every month. C. H. Ump led a pretty sheltered life for a long time, and he really didn't use much of the money. He saved some of the money and didn't even bother to pick up the rest.

Eventually, C. H. Ump married F. R. Ump and they had a gang of little T. H. Umps. What with moving to a split level in the suburbs, 2 cars and a boat, a cleaning lady, and the country club, they began using more and more of uncle T. R. Ump's money. Soon they began using more than they were getting every month, and it was only a question of time before their savings would run out.

In order to keep their savings account from running out, the Umps could pursue either of two options: (1) cut down their consumption to live within their means, or (2) find an alternate money source to supplement their income.

This analogy is unfortunately accurate. Our rich uncle, the sun, is sending us energy-rich photons. For a long time we didn't use much of this energy, and it was stored in the form of petroleum. Now that we have matured technologically and developed some expensive habits, we are using the stored-up energy very rapidly. To keep some stored-up energy in reserve, we are being confronted with the choice of either reducing our consumption of energy, or finding alternate energy sources. Reducing consumption is a solution rejected by many ("Bigger is better; more, more, MORE"), so alternate energy sources are being investigated.

Among these alternate energy sources are direct conversion of solar energy to electricity, geothermal energy, and nuclear reactors.

Direct conversion of solar energy to electricity has been used to provide energy for some satellites, but has never been used on any large-scale projects as yet. If these "solar cells" are ground based, there are problems with clouds blocking the sunlight or long-distance transmission problems if the cells are set up in the desert. If long-distance energy transmission problems are solved, some visionaries predict that artificial satellites will be sent up to capture the sun's energy and "beam it down" to earth.

Geothermal energy uses the high-temperature material from deep in the earth's crust or even the mantle to convert water into steam (like the Old Faithful geyser at Yellowstone National Park). The steam then runs a turbine which stores electrical potential energy the same way coal-fired turbines do. Only a few places have natural access to the hot parts of the earth's interior (like volcanoes), and technologists have not completely solved the problem of drilling holes deep enough and in the right places to take advantage of geothermal energy.

Nuclear reactors are the subject of intense study and debate now because many see nuclear reactors as the best new energy source. There are two different kinds of nuclear reactors, *fission* and *fusion* reactors.

Earlier in this chapter, we discussed the formation of heavy nuclei in stars and found that there is a limit to the size of the nucleus that may be formed. This size limitation comes about because the nuclear force has such an extremely short range that protons on opposite sides of a large nucleus repel each other due to electrostatic repulsion more than they attract each other due to the nuclear force. Suppose we took one of the largest nuclei which occurs in nature, uranium, $^{235}_{92}U$, and shot a neutron into it. If you think of the uranium nucleus as a large liquid drop, you could undoubtedly envision the big drop distorting and then breaking in two. The uranium nucleus is distorted by the neutron which smashes into it, and the distortion causes the protons to get far enough apart so that the electrostatic repulsion

Figure 6.14 *Uranium nucleus undergoing fission*

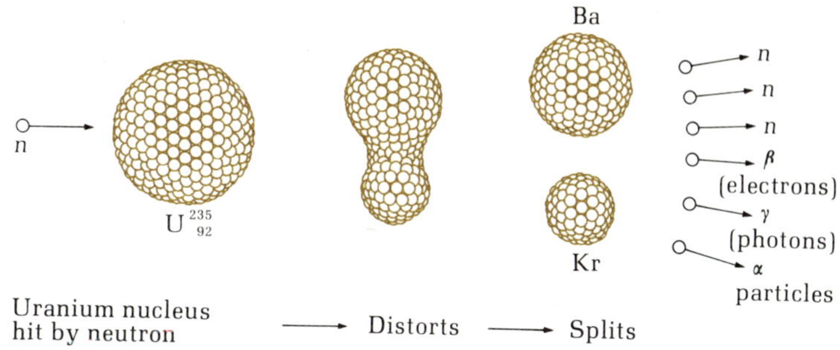

becomes larger than the nuclear attraction, and the large nucleus *splits*. This splitting is termed *fission* because of the similarity to the cell-division process studied in biology. The uranium nucleus doesn't split precisely in two, as shown in Figure 6.14; there are also neutrons, α-particles, β-particles, and γ-rays. (Recall the radioactivity in Chapter 2.)

Since the star which originally made the uranium nucleus had to invest energy to make it, because of its size, when this nucleus splits, forming two smaller nuclei which require *less* potential energy, the additional energy is turned into kinetic energy of the products and γ-rays.

Notice that among the products of this fission reaction are several neutrons. It took a neutron to start the whole fission process, so these neutrons could start further fission processes if they each banged into another uranium nucleus, which would create more free neutrons, which would cause more fissions, and so it goes. If enough uranium nuclei are present (called the *critical mass*), the fission process will work continuously until all the uranium nuclei have undergone fission. This process is called a *chain reaction*.

This chain reaction may be made to occur rapidly by suddenly bringing together two hunks of fissionable material, each of which is individually less than the critical mass, but the total mass when they are brought together is greater than the critical mass. The result? A rapid chain reaction with plenty of high-energy photons emitted and a rapid expansion of the material and its surroundings because of the large kinetic energy of the fission products. In short, a *bomb;* specifically (if you'll pardon the pun) called *A-bombs*.

The chain reaction could be made to occur more slowly if there was something to absorb some of the neutrons so that they would not all cause new fissions. This controlled chain reaction, called a nuclear *reactor*, can be made to produce kinetic energy in manageable quantities, which may be used to turn water into steam and drive a turbine, which is in turn used to power submarines or generate electric energy. Because the petroleum supply is *not* depleted by nuclear reactors, they are a very attractive energy source. However, nuclear reactors do have some disadvantages. *First,* there is the problem of *radioactive wastes.* The fission products are radioactive and must occasionally be removed so that the reactor continues to operate efficiently. What is to be done with these radioactive wastes? In time, their natural radioactivity dissipates as they decay to stable elements; but the α, β, and γ-rays that are emitted as the radioactive elements decay can cause serious damage if they bang into the complex molecular structures of some living organism. *Radiation sickness* is very difficult to treat because it is virtually impossible to get at the radioactive elements which are the source of the problem, because they combine chemically with the organism's molecules, becoming part of the organism. For example, radioactive strontium, $^{90}_{38}Sr$, is directly below calcium, $^{40}_{20}Ca$, in the periodic table, and so it acts chemically just like calcium and gets into the bones.

The *second* difficulty with nuclear power generating plants is the problem of *thermal pollution.* Substantial amounts of cooling water must be pumped through the reactor from nearby rivers, lakes, or ocean to take away excess kinetic energy. When this water is returned to its source, its increased temperature may have a negative effect on aquatic life and, inevitably, man.

Although most people would prefer not to even think about it, a *third* problem of nuclear reactors is *safety.* What if the elaborate control system malfunctioned and a reactor exploded? Ask the Japanese about the effects of a nuclear explosion in a populated area.

An alternative approach to this whole problem of energy generation is a *controlled fusion reactor.* In a sense, we would be creating our own sun because the fusion reaction is simply the proton-proton chain we discussed earlier, which supplies the sun's energy by fusing together four hydrogen nuclei to form one helium nucleus. Again, if this was done very rapidly, it would be a bomb. Naturally enough, we've made a few fusion bombs, called *H-bombs.*

A slow fusion reactor has some distinct advantages over the slow fission reactor:

1. The end product of the fusion reaction poses *no serious disposal problems* because it is not radioactive. It is helium—good for filling balloons.
2. There is no *chain reaction* involved, so we have a greater measure of safety. Unfortunately, the controlled fusion reaction is extremely difficult to achieve. Recall that a high temperature is needed so that the protons will get close enough for the nuclear attractive force to overcome the electric repulsive force. The temperature at which this happens is about 1×10^7 °C. First, how do we get the hydrogen that hot? Secondly, what do we put it in when it gets that hot? Scientists have been working with "magnetic bottles" for a while, and have been investigating high-powered lasers for a short time. Stick around; it may be solved shortly.

In summary, all our rearrangements of solids cost energy. Right now we're going through our saved-up energy at an enormous rate, and we're striving to develop alternate energy sources before our petroleum reserves vanish.

6.4

Solid Junk: 2 to 4 Kilograms per Person per Day

Sometimes we carefully rearrange some solids into different forms, at the expense of prodigious amount of energy, only to find that *some portion* (or *even all*) *of this solid matter has no clearly evident use at the time and place it exists.* This will be our working definition of *solid junk.* We could have been a bit more tasteful and referred to it as refuse, waste, garbage, or trash, but *junk* seems to be the best of all these terms for getting across the notion of having no clearly evident use at this time and place. (People who deal with this stuff every day—janitors, garbagemen, junk dealers, stadium sweepers—have other, more descriptive but less technical, names for it.)

An obvious example of something which is in the wrong place at the wrong time to be useful is the package that a consumable item comes in after the consumable item has been consumed. A beer can is not very useful to you once you have consumed the beer it contained, but it would be useful to the beer can factory because it could be melted down and reused. Similarly, the wrapping paper which gets slightly torn by zealous children in their haste to open their presents has very little use after the gifts have been opened, but could be used at the paper mill as raw material for making more paper.

The fact is that we are generating quite a lot of solid junk these days—two to four kilograms per person per day of it. This may not seem like much until you realize that *a family of four throws away an amount of mass equivalent to more than 2 automobiles every year.* Not just the amount, but also the *variety* is staggering. Besides all the normal *household junk* we put into our garbage disposals, compactors, incinerators, garbage cans, or over the fence into the neighbor's yard, there are also *industrial solid wastes,* which constitute a far greater mass than household wastes. Table 6.1 shows some of the categories of solid waste recognized by the United States government. The amount and variety of junk show how extensive our rearrangement of our planet's surface has become. What we are doing is converting the solid matter from our earth's crust into unusable forms—junk.

Our throw-away society is in the process of learning a painful lesson:

There is no "away" to throw junk.

Try as we may to get rid of it, only two fundamentally different things can happen to junk: *accumulation or recycling.*

Accumulation usually takes one of three forms: *incineration,* an *open dump,* or a *sanitary landfill.*

Incineration involves the formation of small new molecules from large old ones. Because the potential energy decreases when the atoms get closer together in the small new molecules, this potential energy decrease must show up somewhere, according to the conservation of energy, so it shows up as increased kinetic energy (a rise in temperature) of the surroundings. (This is a complicated way of saying that fire is hot!) Many of the usual things which are incinerated are noncrystals which come apart at a low temperature and form CO_2 and H_2O, both of which are in the gaseous state and are carried away into the atmosphere. Unfortunately, some incinerated things form other compounds (ever smell burning rubber?) and we have traded solid pollution for air pollution.

Some new incineration methods have been developed which include secondary combustion chambers to ensure complete combustion (incomplete combustion generates

Table 6.1 *Major Industrial Solid Waste Categories*

Acetylene wastes	Fly ash	Pottery wastes
Agricultural wastes	Food-processing wastes	Precious metals
Aluminum	Foundry wastes	Pulp and paper
Animal-product residues	Fruit wastes	Pyrite cinders and tailings
Antimony	Furniture	Refractory
Asbestos	Germanium	Rice
Ash, cinders, and flue dust	Glass	Rubber
Asphalt	Glass wool	Salt skimmings
Bauxite residue	Gypsum	Sand
Beryllium	Hemp	Seafood
Bismuth	Hydrogen Fluoride slag	Shingles
Brass	Inorganic residues	Sisal
Brewing, distilling, and fermenting wastes	Iron	Slag
Brick plant waste	Lead	Sodium
Bronze	Leather-fabricating and tannery wastes	Starch
Cadmium	Leaves	Stone spalls (chips)
Calcium	Lime	Sugar beets
Carbides	Magnesium	Sugar cane fibers
Carbonaceous shales	Manganese	Sulfur
Chemical wastes	Mica	Tantalum
Chromium	Mineral wool	Tetraethyl lead
Cinders	Molasses	Textiles
Coal	Molybdenum	Tin
Cobalt	Municipal wastes	Titanium
Coffee	Nonferrous scrap	Tobacco
Coke-oven gas residues	Nuts	Tungsten
Copper	Nylon	Uranium
Cotton	Organic wastes	Vanadium
Dairy wastes	Paint	Vegetable wastes
Diamond grinding-wheel dust	Paper	Waste paper
Distilling wastes	Petroleum residues	Wood wastes
Electroplating residues	Photographic paper	Wool
Fermenting wastes	Pickle liquor	Yttrium
Fish	Plastic	Zinc
Flue dust	Poppy	Zircaloy
Fluorine wastes		Zirconium

Source: Bureau of Solid Waste Management, U.S. Department of Health, Education, and Welfare.

carbon monoxide, CO, a deadly gas) and processing of exhaust gases to remove small bits of solids and undesirable gases.

Open dumps are certainly the oldest form of waste disposal, in which junk is compacted, bulldozed, and burned to minimize its volume. Junk dealers often salvage metal from open dumps, and various rodents may salvage organic wastes. Because of the fires and rodents, open dumps are not much favored by communities for esthetic and health reasons.

Sanitary landfills are an improvement over the open dump in that the day's accumulation of junk is covered with a layer of dirt rather than being burned, and the rodents find pickings a good bit slimmer.

The other thing that we do with junk is to *recycle it,* illustrated by D. Ump in Figure 6.15.

Materials which undergo a natural form of recycling are said to be *biodegradable.* A biodegradable substance is one whose molecules are broken down to simpler forms by some naturally occurring process and these simpler molecules enter one of the natural cycles and are reused. Earlier we discussed the energy man obtains from plants and noticed

Figure 6.15 *D. Ump helping recycling*

that CO_2, O_2, and H_2O all participated in the energy flow in a cyclic way. Plants used H_2O and CO_2 as their raw materials and gave off excess oxygen, while man required oxygen and gave off CO_2 and H_2O. Thus, if some large molecule could be broken down by some natural process (like bacterial action) into CO_2, H_2O, or O_2, these molecules could enter the flow of molecules between man and plants and be continually reused and, therefore, recycled.

Large molecules which are biodegradable are nearly always compounds of carbon, and are termed *organic*. Carbon is THE key element in the chemistry of living things, because of its ability to form so many different molecules by combining with itself, hydrogen, and oxygen in a large number of ways. Carbon, $^{12}_{6}C$, has 2 electrons in its innermost energy level and 4 electrons in the outer energy level. Carbon thus needs to obtain 4 electrons to fill its outer energy level or it needs to give away 4 electrons to leave a filled energy level. Because it is difficult to give or take 4 electrons, carbon combines with itself by sharing electrons, forming covalent bonds. Carbon's structure allows it to share one, two, or even three pairs of electrons with other carbon atoms or other atoms, and, as a result, carbon compounds are extremely numerous (more than a million carbon compounds have been identified by organic chemists). Furthermore, carbon compounds are extremely long-chained, branched, and even circular in form and they have molecular masses greater than 10^5 times that of the hydrogen atom. Many of these large molecules are particularly susceptible to being broken down to simpler molecules by natural

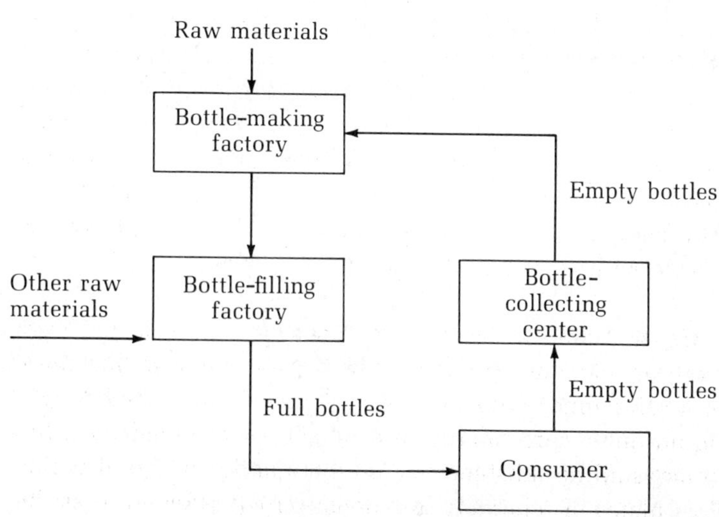

Figure 6.16 *Bottle recycling flowchart*

chemical processes, and these simpler molecules can participate in natural cycles to be reused.

The other type of recycling is artificial recycling that has been set up by man. Glass recycling is a good example of this, as shown in Figure 6.16.

One of the great difficulties with artificial recycling is that the transportation and handling of materials may require a greater energy expenditure than making new manufactured goods and simply allowing unusable old ones to accumulate. Recycling and energy usage must be balanced before a rational strategy is evolved.

Summary

This chapter has dealt with the origin of the mostly solid planet we live on, the arrangement of the solids that comprise the earth, the energy required to rearrange these solids, and finally what happens to solids which have been rearranged to the point in time and space that they are no longer usable—junk.

The Big Bang theory of the origin of the universe is the one currently thought to be the best. It says that about 10^{10} years ago, all the matter and energy in the universe was concentrated into a single, hot, dense fireball. The fireball exploded nonuniformly and some of the hydrogen and helium atoms came together because of gravitational attraction. The protons overcame their electrostatic repulsion and fused together by the nuclear force, formed heavier nuclei, gave off energy, and were called stars. Because of the short-range nature of the nuclear force, there is a limit to the size of the nucleus that is formed, and eventually stars exploded in a nova or supernova, spewing heavy nuclei into space. These heavy elements attracted each other gravitationally and came together eventually to form more stars and planets. This is one theory of how the earth was formed. The Big Bang theory is well regarded because there are several pieces of experimental data which seem to confirm this theory.

Using a density experiment and data from seismic wave experiments, the current model of the earth says that there is a solid inner core of iron and nickel, surrounded by a liquid outer core of iron and nickel, surrounded by a hot noncrystalline substance called the mantle, on top of which float plates of relatively low-density rock referred to as the crust. The plates floating on the mantle move around due to mantle currents and cause earthquakes and continental drift by banging into each other and drifting apart.

Rearrangement of solids costs energy, and the sun is our primary energy source. The sun's energy is continually being stored up by plants, which build large molecules from small ones. Since we are using energy at an extremely rapid rate, we are investigating alternate energy sources, such as direct conversion of solar energy to electricity, geothermal energy, and nuclear reactors, both fission and fusion.

Junk is simply solid matter that has no clearly evident use at the time and place it exists. Our careful rearrangement of the solids on earth, at the cost of great energy, often produces junk at a rate of two to four kilograms per person per day. Industrial solid waste is even more massive, and its variety is staggering. There is no "away" to throw junk. Junk is either accumulated (incineration, open dumps, or sanitary landfills) or else recycled (either naturally with biodegradable materials, or artificially).

Questions

This may be used like a programmed text by referring to the indicated questions while reading the text.

6.1

1. What is the difference between cosmology and astronomy?
2. What is the major difficulty in obtaining astronomical data?
3. What two types of events send cosmologists back to their drawing boards to construct new theories?
4. What is the relationship between the time we have been observing the universe and its lifetime?
5. What force caused the "lumps" to form in the expanding fireball?
6. If two atoms move closer together, what happens to their potential energy? What must then *necessarily* happen to their kinetic energy?
7. Why do protons need so much kinetic energy before they will fuse?
8. What two forces act on the protons around the edges of a star to keep them in equilibrium, rather than either exploding or imploding?

9. What happens to the radiation pressure when the fusion process begins producing elements more massive than iron?
10. When the radiation pressure gets smaller, what happens to the star?
11. Why don't planets get hot as stars do?
12. What experimental evidence is there to support the "Big Bang" theory of the formation of the universe?
13. Explain the "oscillating universe" concept.
14. What experimental difficulties arise in the determination of the total mass of the universe?

6.2

15. How did the earth manage to retain some of the lighter elements when its gravity is so weak?
16. Explain the discrepancy between the density of the surface rocks and the average density of the whole earth.
17. What are the characteristics of the three different types of seismic waves?
18. What do seismic wave records reveal about the interior of the earth?
19. Describe the structure, material, and size of the inner and outer cores, the mantle, and the crust.
20. How could the theoretical model of the earth's structure be used to explain earthquakes?
21. What is thought to cause the relative motion between plates on the earth's crust?
22. Is there any experimental evidence for the existence of currents in the mantle?

6.3

23. In what sense is man's attitude toward his surroundings culturally determined?
24. How is energy involved in our rearrangement of solids?
25. Trace the energy flow for man, petroleum products, and electric power generation.
26. Explain how direct solar conversion and geothermal energy can be used to generate electricity.
27. Explain how large nuclei undergo fission.
28. What is critical mass?
29. Explain how a chain reaction continues.
30. What are the possible disadvantages of electric power generation by nuclear fission plants?
31. What reaction takes place in an H-bomb?

32. What would be the advantages of a fusion reactor over a fission reactor?

6.4

33. What is meant by *junk?*
34. What are the two possible things which can happen to junk?
35. Explain incineration from a molecular standpoint.
36. In what way is a sanitary landfill an improvement over open dumps?
37. Explain biodegradability.
38. What kinds of substances are normally biodegradable?
39. Give an example of a recycling process.

Quotes to Ponder

1. "A space telescope named *Copernicus* rocketed smoothly into a perfect orbit and passed its first critical tests today to give man his clearest view into the universe.

But its principal objectives are hundreds of trillions of miles away. They are great interstellar clouds of gas and dust that scientists believe are the seeds of stars yet unborn and the ashes of those that have come to an explosive end.

The largest telescope, designed by Princeton University, will analyze the ultraviolet star light that has passed through these gas clouds to determine their chemical makeup.

It is believed that cosmic clouds of gas and dust slowly draw together under the influence of gravity until a large amount of gas has condensed, something like a rain drop condenses from an atmospheric cloud. Gravity forces and pressures rise and temperatures finally become hot enough for thermonuclear reactions to occur. The star then turns on.

Copernicus also will examine ultraviolet light from young hot stars which live quickly by astronomical standards. Their life spans are measured in millions of years and during the 20 billion year age of our galaxy many generations of these stars have come and gone." (United Press Release, October, 1972)

2. "Faced with an impending fuel shortage, the United States 'urgently needs a comprehensive energy policy,' says Rawleigh Warner Jr., Chairman of Mobil Oil Corporation.

To make a real dent in rising U.S. energy consumption, he maintains, would require 'stringent rationing and mandatory reduction of living standards, together with zero population growth.'

A rational energy policy would enable the U.S. to add significantly to its domestic energy reserves without degrading the quality of the environment, Warner believes. According to Mobil's projections, the U.S. will have to import over 40 percent of its petroleum supplies by 1980, if present trends continue. Spending some $13 billion on oil imports by 1980 could put an intolerable burden on the U.S. balance of payments. And if foreign supplies were suddenly shut off, U.S. national security could be jeopardized.

Environmentalists are beginning to face up to the energy-gap. They now see that a worsening gas shortage would lead to increased use of alternative fuels that are not as clean-burning." (Martin Skala, "U.S. Energy Policy Urged," C. S. M. News Service, October 8, 1972)

3. "Whisk yourself into the 21st century. Giant windmills whirl offshore. Floating energy plants rely on differences in the ocean's temperature. Overhead, dumbbell-shaped spacecraft capture the power of the sun.

So go the visions of scientists seeking to ease the nation's growing energy crisis as current, environmentally acceptable sources of energy diminish.

The scientists list these two other sources of unconventional energy for possible earlier and relatively wide application:

—Expanded use in the South of roof-top waterheaters fueled by solar power and, in those same sunny states, further use of solar stills for making fresh water from salty, brackish water.

—Generating electricity from two forms of geothermal energy other than the so-called hot-rock type that is deeply locked within the earth.

These more accessible, though geographically less widespread, resources of geothermal energy include steam-yielding geysers and subterranean hot-water systems whose waters can be made to flash to steam if brought to the surface.

The geyser fields of Northern California are the major known domestic source of geyser energy. Power plants of the Pacific Gas and Electric Co. began generating electricity at the field in 1960. The combined output is still modest, much less than the 1000 megawatts a single conventional power plant produces." (Frank Carey, "Scientists Searching for New Sources of Energy," Associated Press News Release, October 12, 1972)

4. "New Mexico, the 'Land of Enchantment,' may become the dumping ground for 27 years of atomic waste.

The Atomic Energy Commission began an inspection last month of salt beds east of Carlsbad to bury waste from nuclear power plants hundreds of feet beneath the earth.

The AEC first plans a pilot disposal site in which waste would be stored in cannisters and buried 2,000 feet deep. The cannisters, transported by rail, would be retrievable for monitoring purposes in the pilot project but not at the permanent facility to be constructed later." ("Site to Dump Atomic Waste Sought," United Press Release, October, 1972)

5. "Up to now the major direct use of solar energy has been the photosynthesis process of green plants. Man's major fuels—coal, gas, and oil—come from decayed plant life containing that energy. As everyone knows, the supply of the fuels is running out. Everyone also knows that alternate power sources are needed. One now gaining new attention is the direct conversion of solar energy to power.

Basically there are two nonphotosynthetic ways to use solar energy: concentrate the rays and use them to heat something and eventually boil water to run a turbine or use the photovoltaic effect. The photovoltaic effect depends on the ability of certain crystals to generate an electrical potential, a voltage, that will drive a current when light shines on them." (Dietrick E. Thomsen, Enterprise Science Service, June, 1972)

6. "Scientists who study California earthquakes will soon look to the sky for clues to what's going on beneath their feet.

Orbiting laser beam reflectors should enable them this summer to begin to measure the creep along the San Andreas Fault over the greatest distances yet surveyed.

The fault marks the boundary between two great plates of earth's crust. That carrying much of the Pacific and part of California is sliding inexorably northwestward past the plate carrying the bulk of North America.

Sometimes and in some places the plates slip quietly and slowly past each other. Elsewhere they may stick, building up strains that eventually release themselves in lesser or greater earthquakes." (Robert C. Cowen, "Orbiting Laser

Beam to Help Measure Earthquake Chances," CSM News Service, July 12, 1972)

7. "Contrary to the impression created by some environmentalists, an aluminum beer can tossed into the woods by a careless hiker will break down and return to nature—eventually.

In the long run—the very long run, according to several materials degradation experts—nature recycles everything.

An aluminum beer can dropped on the forest floor this summer, for example, will probably have broken down into particles of aluminum oxide by the summer of 2472, in the estimation of Dr. Edwin L. Owen, a Pennsylvania State University specialist in corrosion of metals.

One widely used plastic, polyvinyl chloride (PVC), buried in test soil has been found to have a half-life of almost 85 years. That is, after 85 years, half the quantity was gone. In another 85 years, half the remaining quantity will break down.

Thus, a piece of PVC would be about 95 percent degraded by the year 2312 if buried in such soil today. On top of the ground, it would take much longer.

Kammereck said that a problem with the biodegradation of many plastics is that their chemical components are similar to some pesticides. Thus, bacteria may attack a styrofoam coffee cup, but they may release a suicidal amount of toxic chemicals.

Perhaps the least perishable of the commonly discarded materials is glass. Prof. Evelyn C. Marboe, a glass chemist, said a bottle left on the forest floor might not break down for a million years." (Boyce Rensberger, "Everything Gets Recycled in the Long, Long Run," New York Times Service, October, 1972)

8. "Pollution and unsightly litter from bottles, wrappings, and other products now can be overcome by plastic materials which self-destruct or disintegrate upon exposure to sunlight, two scientists said today.

Dr. James E. Guillet of the University of Toronto in Canada showed pictures of a plastic cup gradually disappearing, as though an animal were chewing on it, in 14 days of exposure to daylight. It disintegrated far faster than a paper cup would, Guillet told the annual fall meeting of the American Chemical Society.

Dr. Guillet's approach is to incorporate special chemical

entities, known as ketone groups, which he said act like scissors to chop up long chains of molecules when they become influenced or 'excited' by ultraviolet light. The chopping into smaller units makes the organic material susceptible to bacterial action, and ultimately return to simple materials in the soil." (Alton Blakeslee, "Vanishing Plastic Litter Cure?," AP News Release, August 28, 1972)

9. "Garbage—and its disposal—offers a continuing challenge to scientific ingenuity. Produced in an ever-increasing amount around the world, it is being incinerated, converted into fertilizer, used as landfill, recycled into new products, or dumped carelessly into rivers and seas. Still it piles up. To keep ahead of the accumulating waste, some scientists have even suggested lofting it into perpetual solar orbit or rocketing it into the sun, where it would be consumed by nuclear fires. Now two University of Washington scientists have proposed what may be a much simpler terrestrial solution: burying civilization's debris deep within the earth.

The trick, they explain in Nature, would be to dump packaged waste into the sea off the mouths of fast flowing rivers, which annually wash vast amounts of mud into continental trench areas. Though the garbage would not be drawn far into the earth for many years, it would soon be buried so deep in mud that there should be little danger of pollution during the interim." ("Geophysical Garbage Dump," *Time,* April 5, 1971)

10. "Most scientists now agree that the universe began with the cataclysmic explosion of an extremely dense primordial atom, and that the billions of star-filled galaxies, including the Milky Way, are still rushing outward from the original big bang. The speed of that expansion, astronomers have determined, is decreasing—slowed by the gravitational pull of the galaxies upon each other. What cannot be explained, however, is that the calculated mass of the universe's galaxies is only about one-tenth the amount required to produce that rate of deceleration. Where is the missing mass?

Black Holes. Some scientists think that the answer lies in vast clouds of hydrogen and helium gas discovered in intergalactic space. But others, including cosmologists Alastair B. W. Cameron and James Truran of New York's Yeshiva University, doubt that the total mass of intergalactic gas is sufficient to provide the remaining gravitational pull. Instead, Cameron and Truran suggested at a meeting of the

American Physical Society, the missing mass may be hidden away in a completely invisible form: inside so-called 'black holes' in space.

Although the existence of black holes has never been proved, Cameron and Truran speculate that as much as 90% of the universe's mass may exist in this bizarre form. As they explain their theory, the early universe's mass probably consisted of clusters of huge superstars. These primordial giants, as much as 100 times as massive as the sun, eventually cooled, collapsed, and disappeared." ("Much Ado About Nothing," *Time,* April 5, 1971)

CHAPTER SEVEN

LIQUIDS

Although we have seen some properties of liquids when we melted solids, we're now going to investigate liquids in more detail so we can understand that amazing and plentiful liquid, water. Let's start by asking the same kind of questions we asked about solids in Chapters 2, 3, 4, and 5.

What are liquids made of? How far apart are the molecules in a liquid? What force holds the molecules together in a liquid? What role does energy play in the structure of a liquid? Once we have developed a conceptual model for liquids similar to the conceptual model for solids, we will apply our conceptual model to predict what will

Figure 7.1 *How much wood would a liquid lick if a liquid would lick wood?*

happen when liquids meet up with various kinds of solids, other liquids, and gases.

7.1

A Contrast Between Liquids and Solids

What are liquids made of? Just as we answered this question for solids, ATOMS are the fundamental ingredient of all matter, and they gain or lose electrons to form *ions*, or they share electrons in a covalent bond to form *molecules*. The ions tend to be relatively simple, roughly spherical, while the molecules may vary tremendously in *size*, from only a few atoms to thousands of atoms; in *shape*, from simple and almost spherical to long and rod-like; and in *charge distribution*, from having highly positive and negative regions to being quite symmetrical with only weakly charged regions. (Note: Although liquids may be made up of either ions or molecules, we will often use the term molecules to mean either molecules or ions, just for the sake of convenience.)

How far apart are the ions and molecules in liquids? Recall that we had to use x-ray crystallography to find out how far apart the ions and molecules were in the regular arrays of solid ionic and molecular crystals. X-ray crystallography doesn't work on liquids because liquids aren't regular. However, we can use some common observations to answer this question by contrasting liquids with solids.

Observation #1: Solids generally have a higher density than their liquids. In Chapter 5, the cooling curves all involved solids which sank to the bottom of their liquids when they were being melted.

STUDENT: But, Professor, are *all* liquids less dense than their solids? (See Figure 7.2.)

PROFESSOR G. R. UMP: Don't bother me now; can't you see I'm selling something?

Table 7.1 shows some liquids and solids and their densities.

STUDENT: But Professor, what about water? Ice floats, you know.

PROFESSOR: Yes, I do know. Believe me, it will take some

Figure 7.2 *Professor G. R. Ump explaining density*

Table 7.1 *Liquid and Solid Densities of Some Elements*

ELEMENT	LIQUID DENSITY (g/cc)	SOLID DENSITY (g/cc)
Silver	9.32	10.6
Gold	17.3	19.3
Calcium	1.36	1.55
Potassium	0.82	0.86
Osmium	20.1	22.6
Copper	8.0	9.0

real explaining; but let's postpone that explanation until the next chapter.

STUDENT: All right.

PROFESSOR: Now, since the density of a liquid is *generally* less than its corresponding solid, the molecules must take up more volume and hence are farther apart.

STUDENT: That agrees with what we figured in Chapter 5, also. The liquid had more potential energy than the solid, and it was this potential energy that was given up when the liquid cooled down to form a solid.

PROFESSOR: Yes, during the "flat spot" when the temperature, and thus the average kinetic energy, remained constant, the energy given off to heat up the surroundings came from the decrease in potential energy when the ions or molecules went from the relatively far apart positions in the liquid to the relatively close together positions in the solid.

The molecules are thus farther apart in the liquid than they were in the solid, and we saw in Chapter 3 that the distance between molecules in a solid was approximately 10^{-10} meters. So, the molecules are farther apart than 10^{-10} meters, but how much farther apart? Time for another observation.

Observation #2: An undisturbed liquid surface reflects visible light very smoothly. (Ever see a pond or lake on a calm day? Just like a mirror.) As we saw in Chapter 3, if liquid molecules were nearly as far apart as a visible light wavelength, we'd expect to see diffraction and interference phenomena. We don't see diffraction and interference, so we can conclude that although liquid molecules are farther apart than solid molecules, they aren't *too* much farther apart.

What holds liquids together? Actually, liquids don't hold together as solids do.

Observation #3: Liquids take the shape of whatever container you put them in, whereas solids retain a definite shape. This means that there is no long-range order to the liquid structure as there was in the solid structure. But, on the other hand:

Observation #4: In general, liquids are not easily stretchable or compressible; they tend to retain a fixed volume. This implies that the molecules in a liquid must maintain *some kind* of relationship with each other, although not the long-range, orderly relationship of solids.

Although the details of the interaction between molecules in a liquid aren't clear, we can be sure that the electric force is the force responsible for the interaction, because we can recall from Chapter 4 that it is *the* force which is both attractive and repulsive and universal. For example, notice that liquids seem to come apart under the influence of gravity. (Ever try to hold a cube of liquid water?) This is another indication of the liquid state's lack of long-range order as opposed to the solid state.

SHARP STUDENT: Something just dawned on me! We just finished figuring out that the molecules in a liquid are a little farther apart than the molecules in a solid, and now we're saying that the electric force doesn't hold liquid molecules together as well as solid molecules. That stands to reason, just by looking at Coulomb's law; doesn't it? I seem to recall that the electric force between charges decreased with increasing distance.

EXCITED PROFESSOR: I think you've got something there. Coulomb's law is:

$$F_E = K\frac{Q_1Q_2}{r^2}.$$

As the distance between charges increases, the force decreases, so we can't expect the electric force to hold liquids together as strongly as solids because the molecules in the liquid are farther apart.

Before we try to formulate a model of the liquid state

similar to the solid state model we developed in Chapter 5, let's make a few more observations.

EXPERIMENT 7.1
EVAPORATION

1. *Add a small (about* $\frac{1}{3}$ *of a test tube) measured amount of boiling water to a test tube. Put a stopper on the tube with a thermometer stuck through the stopper so that the end of the thermometer sticks in the water (Figure 7.3). Record the temperature at 30-second intervals for 10 minutes.*

2. *Repeat step #1 (be careful to use the same measured amount of water) using an open test tube.*

3. *Repeat step #1 (again, same measured amount) using an extremely shallow beaker.*

4. *Plot the cooling curves on the same graph (Figure 7.4), starting each with the first temperature below 100°C.*

They should cool at different rates, and the model we develop for liquids must explain why.

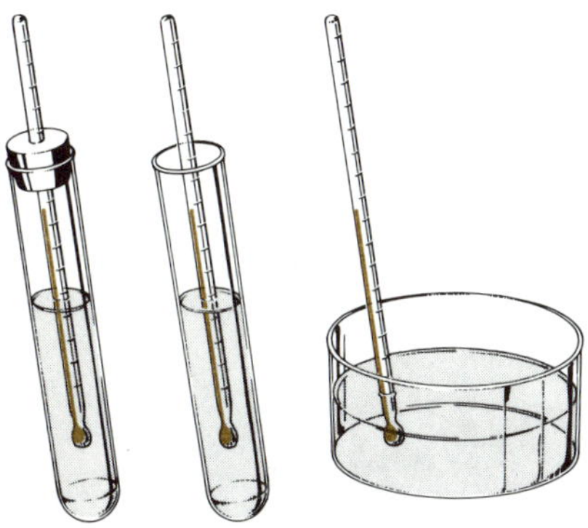

Figure 7.3 *Evaporation experiment setup*

7.2

A Model for Liquids: Entropy

The key to understanding solids, as we saw in Chapter 5, was *energy*—kinetic energy, potential energy, conservation of total energy. The key to understanding liquids will also involve energy.

Recall that when liquids solidified to form crystals, the potential energy diagram for the solid crystal showed a regular series of "wells" in which the ions or molecules vibrated back and forth, like a pendulum or an ice skater suspended between springs (see Figure 7.5).

The liquid, on the other hand, had a higher average potential energy and no regularity in the depth or spacing of the "wells." As the liquid cooled to form a solid, until the solid was *completely* formed, the average kinetic energy (the temperature as we measured it) was the same in the liquid and solid.

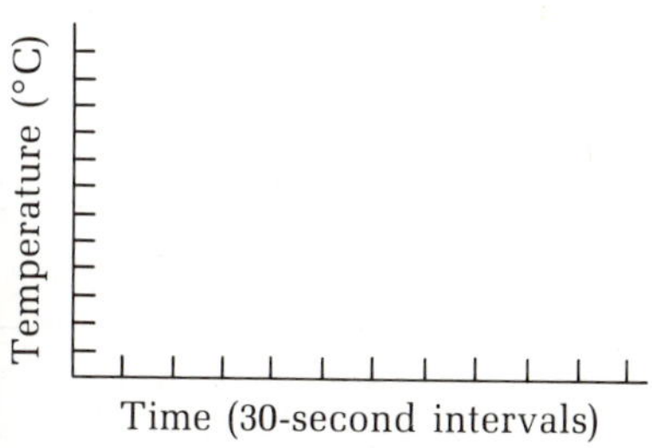

Figure 7.4 *Graphical format for cooling curves*

Right at this point there is a temptation to leap to a very appealing, but FALSE, conclusion. Ask yourself the question:

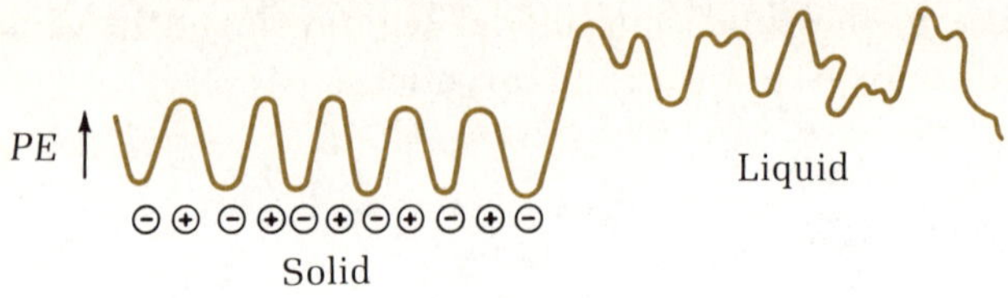

Figure 7.5 *Potential energy diagram for a solid and liquid crystal*

Are the kinetic energies of the molecules and ions all the same or are they different, with some having more kinetic energy and some less?

Since the average kinetic energy of the molecules in the liquid is the same as the average kinetic energy of the molecules in the solid during solidification, it is tempting to conclude that all the molecules have the same kinetic energy. WRONG! Let's try an experiment to shed some light on this question.

EXPERIMENT 7.2
ENTROPY

1. *Put 5 coins in a box, arranged all heads up.*
2. *Close the box, shake vigorously, open the box and count the number of coins which are heads up.*
3. *Repeat step #2 50 times.*
4. *Make a graph showing the number of times each number of heads were up (Figure 7.6).*
5. *Repeat the same process using 10 coins.*
6. *Try to imagine what the graph would look like for 100 coins, 1000 coins, 10^6 coins or more.*

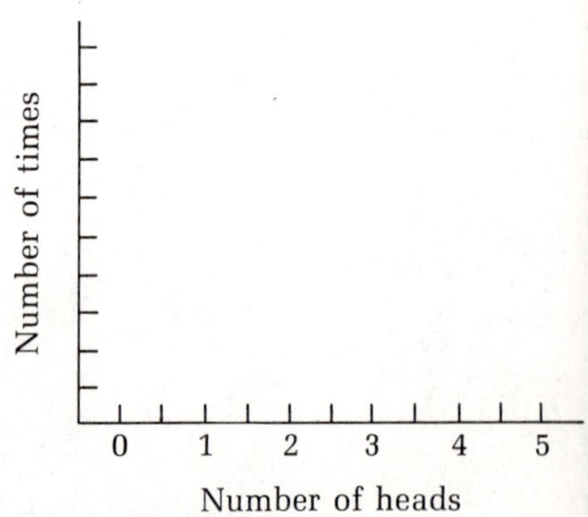

Figure 7.6 *Graphical format for Experiment 7.2*

This experiment demonstrates the concept of increasing randomness (called *entropy*), which we'll discuss in general terms before getting back to the liquid problem.

Let's start by analyzing an even simpler coin problem. If 2 coins are put into a box heads up and shaken, there are only 4 possibilities:

Coin #1	Coin #2	Shake	Coin #1	Coin #2
heads	heads	→	heads	heads
			heads	tails
			tails	heads
			tails	tails

Since there are only these 4 possibilities, you would expect that, over a large number of trials, you would get no heads $\frac{1}{4} = 25\%$ of the time, 1 head $\frac{1}{2} = 50\%$ of the time, and 2 heads $\frac{1}{4} = 25\%$ of the time.

The reason that 1 head comes up more often is simply that there are more ways to get 1 head than zero or 2 heads.

The same is true using more coins. With 5 coins, 2 or 3 heads should predominate; and with 10 coins, 5 heads should come up most often.

If we replace coins with molecules and heads or tails with energy states, then we can identify entropy with the number of ways a particular energy state of the whole set can be achieved.

DAZZLED STUDENT: Wow! Could you go over that again? Maybe an example would help.

PROFESSOR: Delighted to oblige. Suppose we have two molecules (like the two coins) and two different energy states (like heads and tails). If random motion (like shaking the box) can cause either of the molecules to occupy either energy state, then they could both be in the first energy state (like both heads) or one in each energy state (like one heads and one tails, which can be done in two ways) or both be in the second energy state (like no heads). The entropy would be 1 for the case where both molecules are in the first energy state, 1 for the case where both molecules are in the second energy state, and 2 for the case where one molecule is in the first energy state and the other molecule in the second energy state.

STUDENT: Oh, I get it. The reason we did the experiment with the 5 and 10 coins was to show how the tendency to share unevenly between heads or tails or energy states decreases rapidly when you increase the number of coins or molecules.

PROFESSOR: Precisely. The 5 coins should come up all heads only 1/32 of the time (about 3%) and the 10 coins should come up all heads only 1/1024 of the time (about 0.1%). This shows that an uneven distribution is unlikely, to phrase it one way, or that an even distribution is likely. Right here is my chance to put in the *Principle of Increasing Entropy:*

Entropy tends to increase.

That is, distributions tend to go from the less likely and

orderly distributions to the more likely or disorderly ones.

STUDENT: You mean that's all there is to entropy? With such a strange name, I thought it would be a lot more complicated than that. Basically, it says that things will tend toward distributions of higher probability rather than lower probability. That's pretty obvious, isn't it?

PROFESSOR: I'm glad you think so. To further clarify the point, let's try to think of a system of molecules with more than just 2 possible energy states, and see if we can figure out how the energy of the molecules will be distributed among many energy states.

STUDENT: I have some ideas about how the energies of the molecules would be divided among many energy states, but first I'd like to make a suggestion about an experiment to see what would happen with several states rather than just the two, heads and tails. Suppose you used a 3-sided coin or one with even more sides . . . .

PROFESSOR: How about 6 sides? To make this a practical example, let's take a cube, put one dot on one face, two dots on the next, and so forth up to 6 dots. We'll take a pair of these and call them dice (Figure 7.7).

Figure 7.7 *Experimental probability demonstration (dice)*

STUDENT: I know you're going to say there are many more ways for the two dice to show different numbers than to show the same numbers, which again illustrates your point about how it's easier for the energy states to be different because there are more ways for them to be different than the same. But isn't this also illustrated by the fact that you roll more 7's than any other number because there are more ways for the two dice to add up to 7 than any other number?

PROFESSOR: Although I hesitate to suggest experiments involving dice (people might misunderstand our scientific purposes), let's analyze the dice a minute by listing the possibilities.

FIRST DIE	SECOND DIE	SUM
1	1	2
	2	3
	3	4
	4	5
	5	6
	6	7
2	1	3
	2	4

First Die	Second Die	Sum
	3	5
	4	6
	5	7
	6	8
3	1	4
	2	5
	3	6
	4	7
	5	8
	6	9
4	1	5
	2	6
	3	7
	4	8
	5	9
	6	10
5	1	6
	2	7
	3	8
	4	9
	5	10
	6	11
6	1	7
	2	8
	3	9
	4	10
	5	11
	6	12

Out of 36 possibilities, the two dice show the same number only 6 times; so about 83% of the time, the dice show different numbers. (Notice there are six ways to get a 7, five ways to get a 6 or 8, four ways to get a 5 or 9, three ways to get a 4 or 10, two ways to get a 3 or 11, and only one way to get snake eyes, 1 + 1, or boxcars, 6 + 6.)

By the way, mathematicians who work with probability and statistics deal with these kinds of problems often.

Now, the point of all this discussion about entropy was to try to figure out whether the molecules and ions in liquids (and solids, for that matter) are more likely to have all the same kinetic energy or all different kinetic energies.

STUDENT: In view of our discussion, I'd say the *ions or molecules would be most likely to have energies divided up among all the energy states available.*

PROFESSOR: That's correct. The number of energy states available depends on the kind of molecules or ions involved and the temperature; but it is generally true that there are quite a large number of energy states involved, and most probably there are some molecules in each energy state.

Now that we have decided that the ions and molecules have a variety of kinetic energies, let's see if we can sort out the potential energy and come up with a liquid model which explains the last two observations we made and the results of the evaporation experiment (Experiment 7.1).

Because of the variation in kinetic energies, some molecules or ions have *smaller than average* amounts of kinetic energy and consequently get "caught" in the liquid's shallow, irregular potential wells. Enough ions and molecules are "caught" this way to prevent liquids from stretching or shrinking a great deal, or these molecules would be pulled apart too much against the electric attraction or pushed together too much against the electric repulsion. (This explains Observation #4.)

On the other hand, some molecules and ions have *larger than average* kinetic energies; they are mobile and bang around a lot in every direction until they find something to stop them. This is the reason that a liquid takes the shape of its container (thus explaining Observation #3). The high kinetic energy molecules and ions have enough mobility to move around in all directions.

This brings us right to the evaporation experiment. Some of these high kinetic energy molecules have enough energy to escape the liquid. What happens to the temperature of the liquid if the high kinetic energy molecules escape? The liquid's average kinetic energy must go down because it is losing the upper end of the kinetic energy distribution. (See Figure 7.8.)

When the average kinetic energy goes down, so does the temperature. So, evaporation causes the temperature to go down. (Ever wonder why water is carried in canvas bags in the hot sun? The evaporation keeps it cool!)

The experiment was, therefore, something more than a simple cooling curve. The water was not only cooling by giving kinetic energy to its surroundings; it was also losing

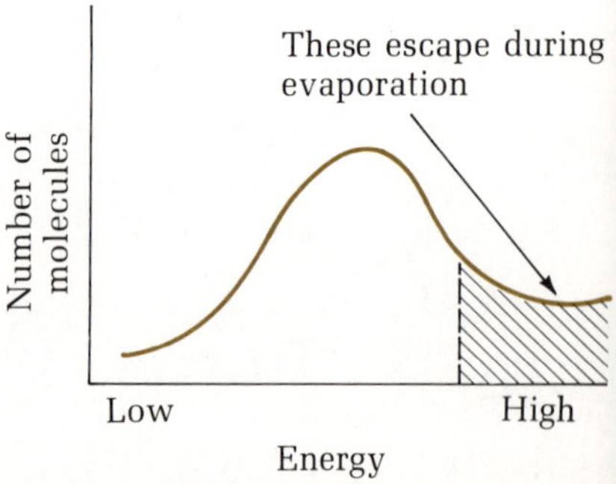

Figure 7.8 *Distribution of energy of molecules in a liquid*

kinetic energy by evaporation. In the closed test tube, the evaporating water molecules had no place to go (we'll talk about saturation presently), so not too much kinetic energy was lost that way. In the open test tube, the molecules could get away somewhat. And in the shallow beaker, the maximum water surface was exposed, allowing a maximum amount of water molecules to escape; so the water in the beaker would have cooled the fastest.

Our conceptual model for a liquid is thus:

A liquid consists of ions or molecules, slightly farther apart than the ions or molecules in a solid. These ions or molecules are in constant random motion and they have a variety of different kinetic energies. Those ions or molecules with low kinetic energy tend to become trapped in the liquid's shallow potential energy wells, giving the liquid *some* structure and keeping it from being pushed together or pulled apart. Those ions with high kinetic energy move around, allowing the liquid to adjust to the shape of its container, and pour. Furthermore, high kinetic energy molecules near the liquid surface may escape, lowering the average kinetic energy and hence the temperature.

Now let's apply this model to see how well it works.

7.3

What Happens When Liquids Meet Other Liquids, Solids, or Gases?

Now that we have a model for liquids—not too detailed a model because scientists don't really understand liquids as well as solids or gases—let's try to figure out what would happen when liquids get together with other liquids, solids, and gases. We'll first conduct an analysis from the molecular point of view, predict what would happen at the observational level, then try it by doing some experiments. (Notice the scientific method at work.)

The more energetic liquid molecules are going to eventually come around and bang into whatever we splash into a liquid, because they are in random motion (as we all are sometimes) and bang into everything. When the liquid molecule gets close to the molecule we toss in, that's when we get down to the nitty-gritty. Try to envision two molecules approaching each other. What characteristics of the molecules determine the behavior of the molecules *after* the encounter? Clearly, the forces involved are electrical, so the *charge distributions* of the molecules involved are all-important.

Recalling our analysis of solids from Chapter 5, the basic building blocks could be *ions*, which are small, simple and highly charged, or *molecules*. Molecules may range from *small* to extremely *large* (e.g., water is a small molecule; some organic molecules, like plastics, can be extremely large), from *simple* to extremely *complicated* (e.g., water is simple; salol is fairly complicated), and from a *strong natural dipole to a weak induced dipole* (e.g., water is a strong natural dipole; oxygen, O_2, is a weak induced dipole).

In summary, the basic building blocks may be categorized in the following way:

Ions: small, strongly charged, simple

- Molecules
  - Small and simple
    - Strong dipole
    - Weak dipole
    - Nonpolar
  - Large
    - Simple shape
      - Strong dipole
      - Weak dipole
      - Nonpolar
    - Complicated shape
      - Strong dipole
      - Weak dipole
      - Nonpolar

To be able to tell which of these categories fits a particular molecule just by looking at its chemical formula would require more chemistry than we have had, so let's look at a few examples to get a feeling for this. (See Figure 7.9.)

Now that we have some raw materials defined, let's play the role of molecular architect and analyze which of these molecules will fit together and which won't, and what *observable* consequences would result.

Solids Meeting Liquids

The crucial point when a solid is immersed in a liquid is whether or not the molecules of the two will "fit" together

Figure 7.9 *Models of several solid, liquid, and gas molecules (illustrated left to right)*

Water: small, simple, strongly polar molecule
Alcohol: small, simple, polar molecule

Acetone: small, simple, weakly polar molecule
Carbon Tetrachloride: small, simple, nonpolar molecule

Sugar: a mixture of many different but similar molecules (one of which is shown) which form a molecular crystal; large, complicated, polar molecule
Soap: a mixture of many different molecules (one of which is shown); large, complicated molecule; polar at one end, nonpolar at the other

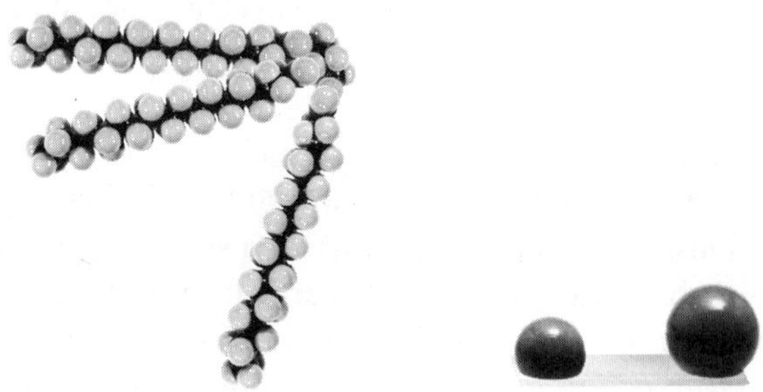

Shortening: a mixture of many different molecules (one of which is shown); large, complicated, nonpolar molecule
Salt: sodium (left) and chloride (right) ions; small, simple, and charged

Oxygen: small, simple, nonpolar molecule
Carbon Dioxide: small, simple, nonpolar molecule

to form a stable unit. The "fit" depends on the electrical properties and size of the molecules involved.

If the Molecules "Fit": The final unit, called a *solvated molecule,* consists of liquid and solid molecules held together by electrical force. However, before it can be formed, the attraction of solid molecules for each other must be broken; the attraction of the liquid molecules for each other must be broken and replaced by the attraction of liquid molecules for solid molecules. Or, in short:

Liquid-Liquid Bonds and Solid-Solid Bonds	Broken and Replaced by	Liquid-Solid Bonds

In terms of energy, enough energy must be supplied for the solid molecules to escape their potential energy wells (Figure 7.10).

The liquid molecules must also escape from their potential energy wells, which are more shallow than those of the solid molecules (Figure 7.11).

Recalling the principle of conservation of energy, the energy necessary to "pry" solid and liquid molecules from their potential wells to form a solvated molecule must be supplied from somewhere. Where must this energy come from? (STOP a moment and think about this question.)

Think for a moment about the solvated molecule. Let's look at a sugar molecule which has met up with some water (Figure 7.12).

Because the sugar molecule is much closer to the water molecules than before they became solvated, this condition is one of *less* potential energy just as a solid has less potential energy because the molecules are closer than in a liquid. Thus, the energy necessary to liberate solid molecules from their attractive solid molecule neighbors, and liquid molecules from their attractive liquid molecule neighbors, comes

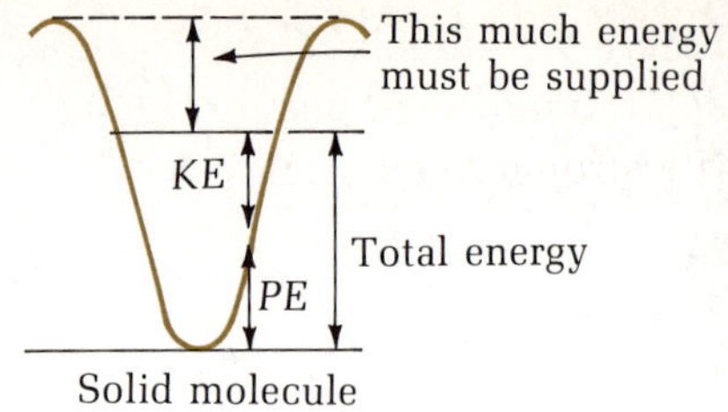

Figure 7.10 *Energy supplied for solid molecule to escape its potential energy well*

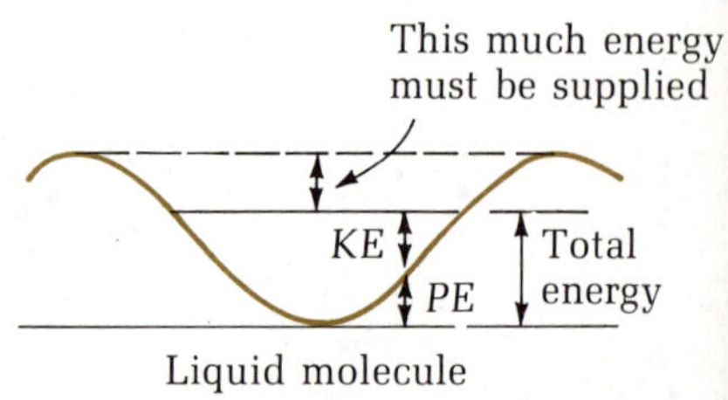

Figure 7.11 *Energy supplied for liquid molecule to escape its potential energy well*

Figure 7.12 *Sugar molecule solvated with water molecules*

Figure 7.13 *Energy balance*

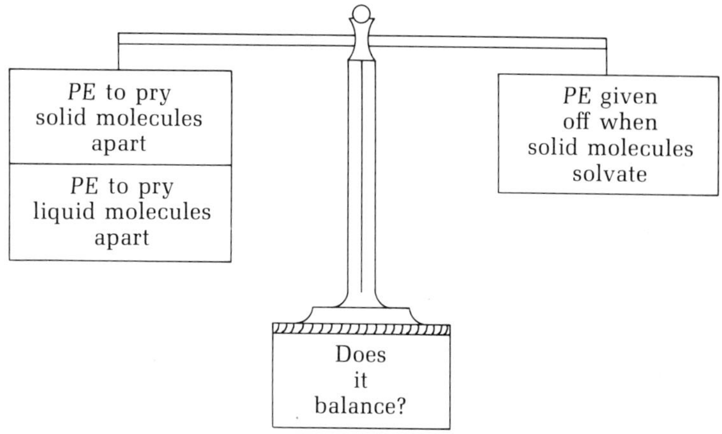

(mostly) from the potential energy released when the solid molecules "fit" together with the liquid molecules and become solvated, or (to use the common term) the solid molecules go into *solution* (see Figure 7.13).

This all seems simple enough, but there are some complications we'll need to look at before we can fully understand solutions.

Complication #1: The potential energy given off when the solid molecule solvates seldom *exactly* balances the *PE* necessary to break the solid-solid and liquid-liquid attractions.

This really isn't too troublesome, because there are always the good old surroundings which will supply energy in time of need. Liquid molecules give up kinetic energy, resulting in lower temperature for the solution until the rest of the surroundings transfer enough kinetic energy to bring the temperature back up. This is called *endothermic* solution formation.

On the other hand, the surroundings are also willing to take any excess potential energy away in the form of increased kinetic energy and hence increased temperature. This is called *exothermic* solution formation.

Complication #2: The potential energy is given off by the solvated molecule *after* the liquid and solid molecules have got together. Where did they get the energy they needed to break their bonds initially so they could get together?

It's true that the molecules can't "borrow" in advance for expected future energy needs. What could be the energy source that would start the solution process? Recall that there is a wide variation in the energies of molecules in the liquid state. If we're willing to wait long enough, a very energetic liquid molecule may come along, with enough energy to knock the solid molecule right out of the grasp of its solid neighbor and form a solvated molecule that way. Or, if you're in a hurry, you can add some kinetic energy to the process and really shake things up. The kinetic energy you add could be in the form of mechanical energy (shake well, agitate, beat, stir) or thermal energy (heat it up to raise the temperature and hence the kinetic energy).

Complication #3: There is a limit to the number of solid molecules that may be solvated.

Often, a number of liquid molecules become involved in solvating a single solid molecule, so it isn't too difficult to imagine a case in which all the liquid molecules that are able are busy solvating solid molecules. Additional solid molecules could not become solvated because all the liquid molecules are busy. This condition is observed, and is called *saturation.* Often, it is possible to dissolve more solid by adding kinetic energy to the liquid. Raising the average kinetic energy means that more liquid molecules are freed from potential wells and hence are available for solvating.

Complication #4: Different solids dissolve in different amounts of the same liquid before the solution is saturated.

Solid and liquid molecules that don't "fit" together too well don't release enough energy when they get together. It may take quite a large number of liquid molecules to solvate them, leading to quick saturation, or slight solubility.

Complication #5: Properties of the liquid change when it contains solvated molecules.

The solvated molecules "tie up" liquid molecules differently from the way they were "tied up" before. As a result, when the temperature drops to the point where the pure liquid would ordinarily freeze, the solvated molecules are still hanging onto the liquid molecules and won't let them form a solid. The temperature must drop still further before a solid is formed, indicating that the solid containing solvated molecules has weaker bonds than the pure solid, because of the additional molecules thrown in. A familiar example of this situation is the action of salt on ice. Since salt is ionic and water is a small, strong dipole, the salt dissolves nicely with the ions, becoming solvated as shown in Figure 7.14.

Figure 7.14 *Sodium (left) and chloride (right) ions solvated by water molecules*

The salt and water solution won't freeze at the normal freezing point of water, so ice is melted by salt. You've probably noticed, however, that when the temperature gets low enough, salt won't work to melt ice; a solid is formed anyway.

If the Molecules Don't "Fit"

Looking at the other extreme, suppose one molecule is highly polar and the other completely symmetric and hence nonpolar. Solid and liquid molecules won't get together to become solvated and release any potential energy, no energy is available to rip apart the solid molecules from their neighbors, and the solid is said to be *insoluble*.

Although you might tend to think about insoluble solids in terms of hunks of steel or rock that sink in liquids with a terrible clunk, some solids float, and others sink so slowly that they might as well be floating. If the pieces of solid are small enough (about 10^{-6} m in size), they are "supported" by being banged into by liquid molecules, and thus settle slowly. The particles are said to be *suspended,* and the solid and liquid mixture is called a *suspension*. If the particles are still smaller (about 10^{-8} m in size), they are called a *colloid* and settle out very slowly. (The study of colloids and suspensions is really much more complicated than this short section implies, but this is enough for now.)

An example of an insoluble solid in a liquid is shortening in water. The nonpolar shortening molecule isn't really able to be solvated by the polar water molecule. Soap, on the other hand, has a polar end and a nonpolar end; so water solvates soap, leaving its nonpolar end sticking out (Figure 7.15).

Starting with a solution of soap in water, can you see how shortening is dissolved? (STOP before you read on!)

The nonpolar end of the solvated soap molecule is able to form bonds with the nonpolar shortening molecules; so the soap acts as a bridge between the water and shortening molecules, allowing them to dissolve.

Figure 7.15 *Water solvating soap molecule*

Liquids Meeting Liquids

Just as was the case with solids, for liquid molecules to be solvated by other liquid molecules, the molecules must basically "fit" together. If they fit, the liquids are said to be *miscible,* and the solution process may be *endothermic* if kinetic energy from the surroundings is required, or *exothermic* if kinetic energy is given to the surroundings. Alcohol and water mix together rather nicely in all proportions, a fact which is demonstrated with great frequency by nonphysical scientists as well as physical scientists. (Ask your local bartender.)

If the liquid molecules don't fit together, then the liquids are said to be *immiscible,* and the liquids will separate with the less dense liquid on top. Examples of immiscible liquids include carbon tetrachloride and water, gasoline and water, and oil and vinegar. If two immiscible liquids are reasonably close in density, kinetic energy may be supplied to cause one liquid to break up into small drops and disperse throughout the other liquid for a time until separation by gravity takes over. One liquid temporarily dispersed in another is called an *emulsion*. Most salad dressings are emulsions, and you must shake well before using to supply enough kinetic energy to break some liquid-liquid molecule bonds to allow the other liquid to get "in," at least temporarily.

A more stable emulsion is milk. Before you get it, it has been thoroughly shaken, and other molecules have been added to increase the settling time.

Gases Meeting Liquids

We haven't discussed gases yet, and we won't until Chapter 9, but it is sufficient for us to realize that they consist of molecules and these molecules are not held together in any way like solids and liquids are. To become solvated, less energy is required because only liquid-liquid bonds need to be broken before the gas molecule may be solvated.

Again, like solids and liquids, those gas molecules that "fit" together best with the liquid molecules dissolve best. For example, although CO_2 is a nonpolar molecule, it has negative "ends," and so it bonds very nicely to one of the positive ends of the water molecule—one of the hydrogen atoms.

Until now, we have stuck to the notion that natural dipole molecules are solvated by liquids which have natural dipolar molecules (or polar molecules, for short), while molecules with no net dipole (that is, nonpolar molecules) are solvated by liquids which consist of nonpolar molecules. The bonds between solvated and liquid molecules are similar to the bonds between molecules in a solid in the sense that they are electrical, but they are generally weaker. This is because of the overall high level of kinetic and potential energy in the liquid, as compared to the solid. The implication in this matching of molecules is that polar molecules *never* solvate nonpolar molecules and vice versa. That is a false implication, and there are a number of species of aquatic life that will testify to its falseness. Oxygen is a nonpolar molecule, yet oxygen dissolves in water. Oxygen doesn't dissolve to any great extent, it's true, but it does dissolve slightly. Just as we discussed the induced dipole bond in Chapter 5, the absence of electrons from the hydrogen end of the water molecule causes the oxygen molecule's electron cloud to become distorted momentarily by attracting the electrons. Momentarily, the $+ - +$ situation exists and there is a bond. The electrons then move on and there is no bond until they return. If this model is true, even the noble gases should be slightly soluble in water (they are), and the ones with more electrons to participate in this induced-charge game should be more soluble than the ones with less electrons (they are). As you may recall, the induced dipole bond is an extremely fragile one, so you would expect increased molecular mobility (higher temperature) to decrease the solubility of oxygen and other nonpolar molecule gases (it does).

EXPERIMENT 7.3

SOLUBILITY OF SOME SOLIDS, LIQUIDS, AND GASES

We have developed a crude model for solubility in terms of molecules "fitting" together, exothermic and endothermic solution formation, and saturation. Simply developing a model isn't going to be sufficient here. We must complete the "scientific method" sequence by making some predictions and then doing some experiments to see how the model works out.

1. *Referring back to Figure 7.9, predict whether each of the solids will be soluble or insoluble in each of the liquids. (MAKE THIS PREDICTION BEFORE YOU PROCEED ANY FURTHER.)*

2. *Perform the experiment by stirring some solid into each liquid, keeping one hand on the beaker to check its temperature. Observe the solid carefully, and note whether some of it disappears (soluble) or doesn't disappear (insoluble), and keep track of "hot" solutions (exothermic) and "cool" solutions (endothermic).*

3. *Choose one of the soluble solid-liquid combinations, continue to stir in more solid, and see if the solution ever reaches a point where it will dissolve no more solid—this is saturation.*

4. *Check the results with your predictions, and try to resolve any disagreements.*

5. *Predict the solubility of the liquids in each other; observe the solubilities and resolve any disagreements between prediction and fact.*

6. *Predict the solubility of oxygen and carbon dioxide in water.*

7. *Determine the solubility of these two colorless gases by starting with water containing no dissolved gases. Introduce the gases into a vessel at atmospheric pressure. Seal the vessel, shake it, and see if the pressure drops, indicating that some gas has gone into solution.*

Liquid Questions

1. Dry cleaning isn't really *dry*. Liquids other than water are used, such as carbon tetrachloride, CCl_4. Use the molecular structure of fibers, oily dirt, water, and CCl_4 to explain why some garments must be "dry" cleaned.

2. Use the properties of water and glass molecules (glass consists of a mixture of large molecules with positive and negative regions) to explain why water drops stick to glass, causing water spots. Speculate about the properties of the molecules in some dishwasher detergents and additives which reduce water spotting.

3. Suppose the energy diagram for water looked like that shown in Figure 7.16. What would this diagram look like if the temperature were raised? How would this affect the

Figure 7.16 *Energy diagram for water*

molecules' mobility and hence the ease with which water pours?

4. Draw an energy diagram similar to Figure 7.16 for liquid molasses at room temperature, based on your knowledge of the difference of physical properties between water and molasses.

5. Explain how liquid lubricants work, particularly in terms of the shape of the molecules of the surfaces to be lubricated and the lubricant. In terms of kinetic energy of molecules, why are liquids usually better lubricants than solids?

6. Sometimes there are chemical reactions between solids and liquids, with new molecules being formed. Explain what is occurring at the molecular level.

Summary

In contrasting solids and liquids, we asked the same series of questions about liquids that we asked about solids in Chapters 2 through 5. What are liquids made of? Ions and molecules. How far apart are the ions and molecules that make up liquids? Observation #1: Solids generally have a higher density than their liquids; Observation #2: Liquids have surfaces smooth enough to reflect visible light without diffraction or interference effects. Conclusion: The ions and molecules are farther apart in a liquid, hence the lower density, but they are not far enough apart to cause diffraction and interference phenomena with visible light. What holds liquids together? Observation #3: Liquids take the shape of their container; Observation #4: Liquids are not easily stretched or compressed. The electric force still works, but not as strongly as in solids because the distance between molecules is greater in a liquid, and the electric force gets smaller with increasing distance.

An evaporation experiment showed that boiling water cooled off fastest in a shallow container. Our conceptual model of a liquid must be able to explain evaporation.

The kinetic energies of the molecules in a liquid are not all equal. Some molecules have large kinetic energies, most have nearly the average, some molecules have small kinetic energies. The principle of increasing entropy says that an extremely even distribution of energies is very unlikely and such a distribution would change rapidly to an uneven distribution. Because molecules have a variety of kinetic energies, those with large amounts of kinetic energy bang around, causing the liquid to flow and take the shape of its container; while those molecules with a small amount of kinetic energy become trapped in the liquid's shallow potential wells, giving it a structure, preventing the liquid from being stretched or compressed. The variety of kinetic energies also explains the evaporation experiment, because high kinetic energy molecules near the surface escape, lowering the average kinetic energy and hence the temperature.

We analyzed what happens when liquids meet up with solids on the basis of the characteristics of the molecules which make up the solid and liquid—the molecular size, shape, and electrical charge distribution. For solid molecules and liquid molecules which "fit" together, the bonds which hold the liquid molecules together and the bonds which hold the solid molecules together must be broken, and new bonds between the liquid molecule and solid molecule must be formed. The energy necessary to break these bonds comes from the lesser potential energy of the solid molecule-liquid molecule combination (called a solvated molecule). The energy released by the new bond formation seldom exactly balances the energies needed to break the old liquid and solid bonds, so the surroundings supply the additional, or absorb the excess, energy. Initially, the first solid molecules are solvated by the high kinetic energy liquid molecules. Sometimes, more than one liquid molecule may be required to solvate a given solid molecule, or else all the "free" molecules are used up such that no more solid molecules may be solvated. This is called saturation. Because solvating solid molecules "ties up" liquid molecules, some properties of the liquid change when solids are in solution. A familiar example is the use of salt to lower the freezing temperature of water.

If the liquid and solid molecules don't "fit," the solid is insoluble, but it may form a suspension or colloid depending on the size of the solid molecule group.

If liquid molecules "fit" other liquid molecules, the liquids are said to be miscible, but if they don't fit, they are immiscible and may form an emulsion.

Gases may be slightly soluble in liquids even if the molecules don't fit too well, because there are no bonds to break between gas molecules.

Using our model for liquids and looking at molecular models of a number of molecules, some solubilities were predicted and an experiment was run to determine how well the model's predictions fit reality.

Questions

This may be used like a programmed text by referring to the indicated questions while reading the text.

7.1

1. What two common observations helped us answer the question about how far apart the atoms, molecules, and ions are in liquids?
2. Do two charged objects have more electrical potential energy when they are far apart or close together?
3. If the molecules in liquids were as far apart as a wavelength of visible light, what would reflections from a calm liquid surface look like?
4. What are the structural implications of the observations about liquid's ability to assume shapes of containers yet maintain a reasonably constant volume?

7.2

5. Why is it tempting to conclude that the kinetic energies of all molecules and ions in a liquid are all the same?
6. What was the point of the analysis of the coins?
7. How does entropy relate to the coin experiment?
8. What are the chances for a whole group of molecules to all have the same kinetic energy?
9. What is the principle of increasing entropy?
10. What was the point of the dice analysis?
11. What would be the most likely distribution of kinetic energies of a group of molecules—all the same or some different?
12. What happens to molecules with much *less* than average kinetic energy?
13. What happens to molecules with much *more* than average kinetic energy?
14. How can we use the kinetic energy distribution to explain evaporation?

7.3

15. Thinking of two molecules about to interact, what characteristics of the molecules determine their behavior after the interaction?
16. What does the "fit" between molecules depend on?
17. What is a "solvated molecule?"
18. What energies are involved in the dissolving of solids by liquids?

19. What happens to the excess or deficiency of energy if the potential energies necessary to break the solid-solid and liquid-liquid bonds don't balance the potential energy decrease of the liquid-solid bond?
20. What is the difference between endothermic and exothermic solution formations?
21. Where does the original bond-breaking energy come from?
22. Explain what is meant by saturation.
23. What is the structural reason for the different amounts of different solids dissolving in the same liquid?
24. Why do solutions have different physical properties from those of pure liquids?
25. What structural reason accounts for insolubility?
26. What is a suspension?
27. What is a colloid?
28. Explain the action of soap on the basis of its molecular structure.
29. What molecular structural properties of liquids lead to miscibility vs. immiscibility?
30. What is an emulsion?
31. What molecular structural properties of gases determine their degree of solubility in liquids?
32. Explain how oxygen, a symmetric molecule, dissolves in water, a polar molecule.
33. What physical property of oxygen dissolved in water demonstrates the fragility of this bond?

Quotes to Ponder

1. "There are still many fascinating problems and unanswered questions at all frontiers of science. We are not yet skillful enough to deal with complexity in nature. Even the structure of liquids is not well understood. No physicist would have predicted the existence of a liquid state from our present knowledge of atomic properties." (Victor F. Weisskopf, noted theoretical physicist, from the essay "The Significance of Science," reprinted in the book *Physics in the Twentieth Century,* MIT Press, Cambridge, Massachusetts, 1972)

CHAPTER EIGHT

WATER

Now that we know a few things about liquids and their interactions with other forms of matter, let's investigate that strange but plentiful liquid, *water*. We'll see the ways in which water is strange—as I. C. Ump discovers in Figure 8.1—and how these strange properties can be explained by water's molecular structure. Natural water contains plenty of things other than pure H_2O; and we'll examine how these things get into the water. Man has influenced water in a number of ways; and we'll examine the effects of man's influence and how these effects may be modified.

Figure 8.1 *I. C. Ump finds some strange water*

8.1

Strange Stuff, Water

How can water be strange? It covers more area of our earth than does solid land; it comes out of our faucets, flows in the rivers and lakes, falls from the sky, and even makes up the major part of the human body. How can a substance this plentiful and all-pervasive be considered strange? Let's look at some properties of water that make water "strange," and try to explain them in terms of what we know about the molecular structure of solids and liquids.

Strange Property #1

Water "should" be a gas at ordinary temperatures.

Suppose we had never seen or analyzed any water, and we wanted to predict some of its properties. Recalling our dealings with the periodic table (page 40) in Chapter 2, elements in the same vertical column have similar chemical properties; so let's take the element directly below oxygen—sulfur. Sulfur is element number 16, so it has 16 electrons arranged in a 2 + 8 + 6 fashion, and since its atomic mass is 32, it has 16 protons and 16 neutrons in its nucleus. Because sulfur is similar to oxygen, it should be able to form covalent bonds with 2 hydrogens as oxygen does, and form H_2S.

H_2S is a molecule familiar to many chemistry students and some chicken farmers. It is a gas, and its characteristic odor is that of *rotten eggs*. (Don't be alarmed; no experiments are scheduled for this topic.) H_2S boils at $-61°$ C and freezes at $-82°$ C, so it is a gas at normal temperatures.

Below sulfur in the periodic table are selenium (Se) and tellurium (Te). These elements also combine with hydrogen to form H_2Se and H_2Te. Their boiling and freezing points are given in Table 8.1.

Let's use this table to predict the properties of H_2O. Since all these substances are gases at normal temperatures, we

Table 8.1 *Boiling and Freezing Points of Some Water-like Molecules*

Substance	Molecular Mass (Atomic mass units)	Boiling Point (°C)	Freezing Point (°C)	State at Room Temperature (~20° C)
H_2S	34	−61	−86	Gas
H_2Se	81	−42	−64	Gas
H_2Te	130	− 2	−49	Gas

would expect water to be a gas also. Furthermore, since the boiling and freezing temperatures are increasing with increasing molecular mass, we would expect H_2O to freeze below −86° C and boil below −61° C. If we had really never seen or analyzed water, we would be amazed by the following facts: it is a liquid at normal temperatures, freezes at 0° C, and boils at 100° C. Where did our predictions go wrong? We'd better get back to basics and see what all these molecules look like (see Figure 8.2).

The molecules are all "bent;" that is, the hydrogens are not arranged in a straight line but are at about 90° angles to each other. H_2S, H_2Se, and H_2Te are all rather weak dipoles, however, because the single proton and electron of the hydrogens don't change the charge distribution of the S, Se, and Te molecules very much. Because these dipoles are so weak, *induced dipole bonds* are important to the formation of solid and liquid H_2S, H_2Se, and H_2Te. Notice that since Te has the most electrons, it can form the "best" induced dipole bonds; so it should remain solid and liquid to higher temperatures than H_2S and H_2Se. Thus, H_2O would be a gas at normal temperatures, with low boiling and freezing points, if it had to rely on the rather fragile induced dipole bonds to form solids and liquids. Oxygen has only 8 electrons, however, so oxygen's nucleus is "shielded" by

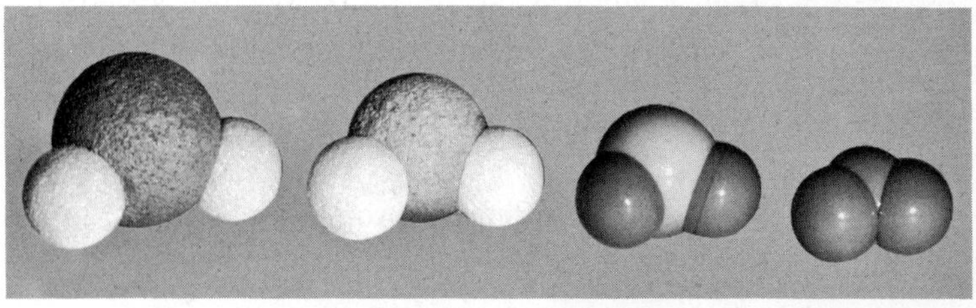

Figure 8.2 *Representations of H_2Te, H_2Se, H_2S, and H_2O molecules (left to right)*

fewer electrons than S, Se, or Te, so oxygen hangs onto the electrons the hydrogens share with it rather tightly, making the overall molecule a strong natural dipole and exposing the hydrogen's bare proton nucleus to the world. Because of the strong natural dipole and the exposed protons of the hydrogen atoms, the H_2O molecule bonds to other H_2O molecules much more strongly than H_2S, H_2Se, or H_2Te molecules bond to each other. The fundamental reason that H_2O is *not* a gas at normal temperatures, like H_2S, H_2Se, and H_2Te, is that the H_2O molecule is a much stronger natural dipole than the others because of oxygen's fewer electrons. This causes a stronger covalent bond between the hydrogen and oxygen atoms, exposing hydrogen's proton.

Strange Property #2

Solid water "should" sink in liquid water.

Again casting ourselves in the role of being unfamiliar with water, if we set ourselves the task of predicting whether solid water would have a greater or lesser density than liquid water, we might look up the densities of solids and liquids in a handbook or encyclopedia. We would find that well over 90% of all substances have a greater density in the solid state than in the liquid state. (Some exceptions: bismuth, gallium, germanium, and silicon.)

If we made our prediction about water based on the densities of other solids and liquids, we would suspect that ice would have a greater density than liquid water, so ice would sink. As the student pointed out to Professor G. R. Ump in Figure 7.2, ice floats. Various forms of aquatic life are well aware of this property, but we need to get down to the molecular level to explain it. Let's look at an ice crystal to see how the water molecules are arranged.

Because of the rather peculiar shape of water molecules, ice crystals have a basically tetrahedral structure. This is an open structure, as Figure 8.3 shows. There are some scientists who theorize that some of the tetrahedral structure is retained in liquid water, but only in fragments, with other water molecules filling in many of the open spaces. Thus, the open structure of the ice crystal, necessitated by the shape of the water molecule, causes the molecules to be farther apart in the solid than in the liquid, leading to a smaller solid density than liquid density.

Figure 8.3 *Ice crystal showing water molecules*

Strange Property #3

Water "should" be capable of dissolving only a few substances.

Most liquids have a rather limited number of substances that can be dissolved by them, and the substances that can be dissolved are mostly ones with molecules similar to the liquid molecules. Water dissolves not only small natural dipole molecules like itself, but it also dissolves many ionic materials, many large natural dipole molecules, and even some induced dipole molecular substances. In fact, water is regarded as the closest liquid to that ideal concept: *the universal solvent.* (If there were such a thing, what would you store it in?)

To explain this unusual property of water to dissolve many substances, we must again look at the unique structure of the water molecule. Its strong natural dipole and positively charged protons sticking out its ends are strong enough to rip apart many ionically bonded crystal structures (like salt) as well as solvating most natural dipole bonded structures (like sugar), yet it is not *completely* ionic so it can form induced dipole bonds with symmetric molecules (like oxygen).

Water has other unusual properties, but these should be sufficient to convince us that water is, in a very real sense, strange.

8.2

What Sort of Natural Water Does the Earth Have? Some Oceanography

Now that we've established water's strangeness or unusualness, let's look at the other side of the coin and see it in all its usualness. Natural waters are anything but "pure" H_2O. Because of water's almost unique capacity to dissolve a wide range of substances, "pure" water doesn't stay "pure" for long.

Water "picks up" substances from the atmosphere, from the solid parts of the earth, and from living organisms and their decomposition products. The substances that are "picked up" may be carried along in suspension (e.g., dust, soil, bacteria, etc.) or as colloids (e.g., fine dust, fine soil, viruses, etc.) or in solution (e.g., CO_2, SO_2, N_2, NaCl, K^+, NH_4^+, NO_3^-, SO_4^{-2}, etc.).

Liquid water, with its suspended, colloidal, and dissolved particles, constitutes over 70% of the earth's surface. The areas where liquid water collects are called various names: oceans, seas, lakes, ponds, rivers, streams, creeks, swamps, lagoons, etc. Let's look just a little more closely at rivers, lakes, and oceans.

A *river* is a body of water flowing from higher points to lower points because of gravitational force on the water. Soil particles and naturally occurring salts are suspended or dissolved and carried along, carving out some pretty good-sized chunks of solid earth in some cases.

A *lake* is often the recipient of water flowing from a river, and is a relatively still body of water compared to a river. Although there are exceptions, lakes are generally fairly small, shallow, and have been formed fairly recently, compared to the oceans. The water in lakes often contains some dissolved natural material, but some lakes are quite still and suspended particles often have a long enough time to settle out.

An *ocean* is generally a large, deep body of water formed a long time ago. Oceans are large enough bodies of water that they respond to the gravitational pull of the sun and the moon by a large-scale movement called *tides*. Another mechanism for large-scale water movement includes *ocean currents,* involving an interaction between the ocean, the atmosphere, and the earth's rotation. There are several of these currents, and they flow in the same fixed directions year after year; in a sense they are more permanent than the rivers running through rock and soil which tend to wander a bit. The Gulf Stream flows past Florida, carrying many times more water than all the rivers draining all the continents.

The size and depths of oceans are directly related to the drifting continents and the thin spots in the earth's crust discussed in Chapter 6. *Oceanography* includes the study of ocean depths, water temperature, the amount and kinds of suspended and dissolved substances, their variation with depth, living organisms in the water and on the bottom, currents, wave and tide patterns, and the composition of the ocean bottom and the crust below the bottom. If oceanog-

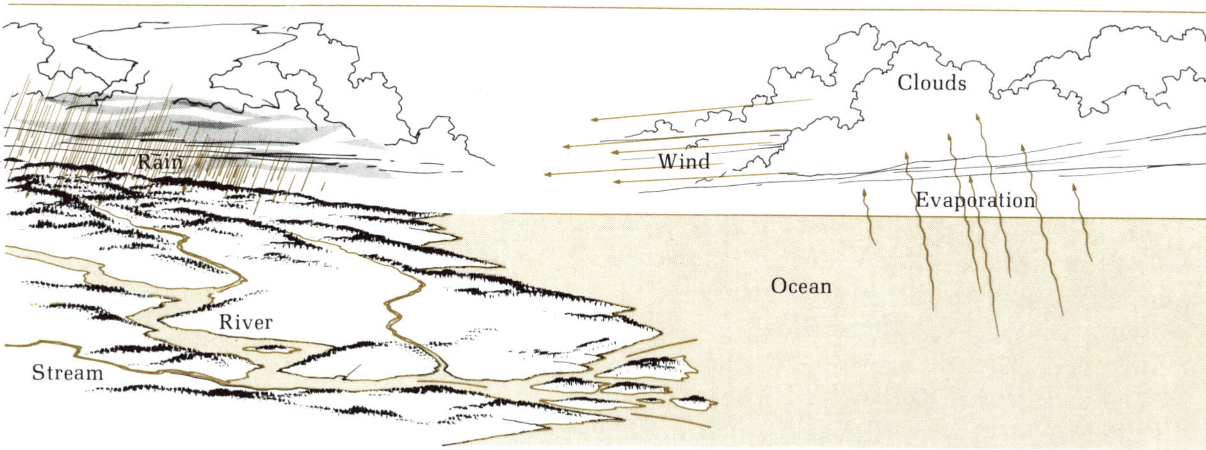

Figure 8.4 *Water cycle*

raphy sounds like a pretty comprehensive job, that's because it is.

The chief distinction between lakes and oceans, besides size and depth, is that oceans have a much larger amount of dissolved materials, particularly salts. To see why salts accumulate in the ocean, we need to look at the whole water cycle. (See Figure 8.4.)

Water flows down the river to the ocean, carrying dissolved, colloidal, and suspended materials along with it. The water in the ocean evaporates because the most energetic liquid molecules have enough energy to escape the liquid bonds. The dissolved salts, however, cannot escape because solvated molecules have considerably more mass; so although they may have plenty of kinetic energy, $1/2\ mv^2$, they don't have enough velocity to escape because their mass is large. Thus, the water molecules escape, but the dissolved salts linger on. The gaseous H_2O in the form of clouds is blown about by winds and weather (see Chapter 9) and eventually comes together enough to fall as liquid water in the form of rain. The rain hits the ground and flows downhill to the river, dissolving salts and suspending solids as it goes. Water flows down the river to the ocean . . . [Return to top of this paragraph and repeat this cycle (recycle) as often as you wish. Water does it all the time.]

Besides liquid water, which we have just discussed, and gaseous water, which we will discuss a bit more in the next chapter, solid water also plays a role in our environment. Certainly, snow and ice are good for skiing, bobsledding, skating, and all that, but solid water had a large role in shaping the present contours of all of Canada and some of the northern United States. Huge ice masses, called *glaciers*, have made at least four major advances and retreats over the northern United States over the last 10^6 years. Because of the very large weight of glaciers, the crystal structure of the ice at the bottom of the glacier is prevented from solidifying completely and the glacier slides along, carrying rocks, breaking down ridges, and generally altering the terrain by its presence. There are a number of characteristic formations visible in the northern states that were caused by the presence of glaciers. Furthermore, their heavy weight pushed the "plate" on which the glacier rested far enough into the mantle that the land level dropped and the ocean covered part of the northern states. When the glaciers retreated to the north, the plate rose, the ocean retreated to the east and west, and the land became dry. Many marine skeletons from this era may be found in the northern states as evidence that this area was once under the ocean.

8.3

What Has Man Been Doing to Natural Waters?

Man's tendency to "use" (manipulate, modify, change, influence, degrade, etc.—choose your favorite word) his natural surroundings to suit his own purposes is clearly evident in his dealings with water. A number of man's *uses* for water may be identified:

Urban Water Supplies: Marvelously complex and intricate systems of plumbing have been constructed at great cost to transport water from rivers, lakes, ponds, wells, etc., into almost every habitable building. Once the water is inside our buildings, we use it for air cooling, washing, cooking, laundry, gardening, showering, bathing, making ice cubes, and sometimes even drinking.

Waste Disposal: Because water dissolves, suspends so many things, and flows "away," it is very handy to "get rid of" various unwanted substances like human and animal wastes (see S. Ump's comment in Figure 8.5) that harbor harmful organisms that have caused terrible health hazards in the past. Furthermore, too much rainfall is also a type of waste, as it is also channeled into the waste disposal system known as the sewer system.

Figure 8.5 *S. Ump's waste-disposal system*

Energy Generation: Flowing water is used to drive turbines which generate electrical energy, as discussed earlier in Chapter 6. Although only about 5% of the total electrical energy is generated using flowing water, some communities rely solely on this type of energy generation. Petroleum burning and nuclear energy generating plants also use water as a cooling agent and as an ash carrier.

Irrigation: Since water is essential to agriculture, areas with inadequate rainfall must have additional water supplied to grow good crops. To this end, dams, canals, and other water distribution systems are built to divert running water to low rainfall areas.

Navigation: Once water was the world's most important transportation medium, but with the growth of population and the modern emphasis on speed, ships tend to carry mostly items which are too large for air or land transport or for which speed of delivery is not vital.

Recreation and Esthetics: Swimming, surfing, boating, sailing, water-skiing, fishing, and canoeing are all leisure-time activities which involve water; and if all predictions of increased leisure time can be believed, these activities will become more important in the future. Waterfront building sites have been traditionally valuable, and ponds, lakes, pools, and fountains are often part of modern developments. Furthermore, many bodies of water are preserved in their natural states for people to observe, study, experiment in, or just plain enjoy. This is normally referred to as conservation.

Industrial Uses: Water is used in many chemical and manufacturing processes (e.g., pulp and paper production and food processing) to flush away wastes and to dissipate excess heat.

Unfortunately, these uses often conflict. For example, one community's waste-disposal system may drain into a river which is another community's water supply; or water which has been warmed or added to by industry spoils the recreational and conservation uses with a large fish kill. The term normally applied to the cause of these and other cases is *water pollution*. Strictly speaking, pollution is any departure from purity. Yet we are now (perhaps painfully) aware of water's ability to dissolve or suspend so many substances that pure water is simply never found. Let's use a working definition of *polluted water* as *water which contains substances which render the water unsuitable for an intended use*.

Let's look at some of these polluting substances, the way they pollute the water, and the way in which they are generated.

Oxygen-Demanding Wastes

Recall that oxygen is somewhat soluble in water. Fish and other aquatic life depend on the oxygen dissolved in water to live. (Have you ever noticed how people whose hobby is watching or raising fish usually have pumps bubbling air through their big aquariums?) They need plenty of oxygen, and if the dissolved oxygen (often abbreviated DO in scientific literature) level falls very much from its saturation level, aquatic life is in trouble, as depicted in Figure 8.6. Domestic and animal sewage, food processing plant wastes, and paper-mill and hide-tanning wastes all contain organic compounds, and hence carbon. With the help of bacteria, the carbon combines with the oxygen dissolved in the water to form CO_2. If much carbon must be oxidized in this way, there results a competition between the aquatic life and the carbon in the wastes for the dissolved oxygen.

Not all organic wastes undergo decomposition by bacterial action, so measuring the total organic content of water wouldn't give a true picture of the oxygen depletion. A sample of the water containing organic wastes is kept at a standard temperature for a standard length of time, and the

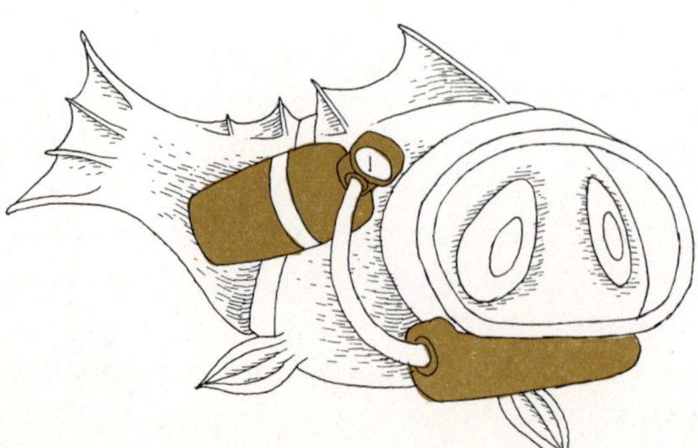

Figure 8.6 *J. Ump feels the effects of low DO level*

amount of oxygen used to oxidize the carbon is measured. This is referred to as the *biochemical oxygen demand* (BOD). A very high BOD can deplete water's dissolved oxygen faster than it can be replaced by oxygen from the air; eventually the lack of oxygen kills the bacteria that would be oxidizing the carbon if oxygen were present. These oxidizing bacteria, called *aerobic* bacteria, are not the only ones present in water. Other bacteria which do not require oxygen, called *anaerobic* bacteria, begin to work if the oxygen level falls too low, and they produce methane gas (CH_4), amines, and H_2S. Methane is flammable, amines have a fishy smell, and H_2S is rotten-egg gas. Now you can begin to see why fish die, rivers catch fire, and terrible-smelling gases are generated when great quantities of oxygen-demanding wastes are dumped into the waters.

Disease-Causing Agents

Human and animal wastes may contain harmful bacteria, particularly bacteria which cause intestinal infections, polio, or infectious hepatitis. To determine the presence of these harmful bacteria, waste water could be tested for each disease-causing strain. Because these tests would require much time, and something must be done with the water rather promptly, a simpler test has been evolved. Besides harmful bacteria, human and animal wastes also contain many perfectly harmless bacteria. One of these harmless varieties, *Escherichia coli* (usually abbreviated *E. coli*), may be detected quite readily. Since these bacteria normally live in the human or animal intestine, they don't survive well in waste water and die off fairly rapidly. The presence of the harmless *E. coli* indicates the "freshness" of the waste and points to a probability of live harmful bacteria also.

Plant Nutrients

As water drains into a lake carrying suspended solids and dissolved minerals, the solids eventually sink to the bottom while the minerals nourish and encourage the growth of aquatic plants (e.g., algae, water hyacinth, etc.). As the plants go through their normal life cycle, some die and sink to the bottom. As they decay, they use up dissolved oxygen, while the floating live plants cut down the water surface area and reduce the opportunity for atmospheric oxygen to dissolve in the water, eventually causing the death of aquatic animals which require oxygen. The bottom keeps rising (or saying it the other way, the lake gets shallower) as it accumulates suspended solids and dead plant carcasses. The lake becomes a swamp, and the swamp eventually becomes a meadow. This is a natural process, called *eutrophication*, and it normally takes thousands of years to accomplish.

The aquatic plants, which are the critical links in this process, require a wide variety of elements (e.g., carbon, hydrogen, oxygen, nitrogen, phosphorus, sulfur, potassium, etc.) in various amounts for proper growth. Natural runoff rainwater contains rather small amounts of some of these elements, and this lack of key elements normally limits the growth of aquatic plants. However, with our modern tendency for speed, we've unwittingly stumbled onto a way to accelerate the eutrophication process. All we have to do is to supply the key elements required by the aquatic plants which are missing from normal rainwater. Depending on the specific plants involved and the chemistry of the lakes, different elements may be the key elements. Some situations have been noted in which nitrogen was the missing element for aquatic plant growth until it was provided in runoff waters by artificial fertilizers used in agriculture. Other cases involved a phosphorus deficiency until it was supplied in the form of phosphates used in modern detergents. Detergents' main active ingredient is a molecule similar to soap in that it has a charged end which sticks to water and an uncharged end which sticks to grease and other symmetric molecules. Dissolved salts in water (referred to as "hard" water) form ions and cause lots of trouble by combining with the charged end of soap molecules, forming insoluble molecules which promptly precipitate out of the solution (e.g., the tattletale gray "bathtub ring"). The molecular architects (chemists) who design detergents added an ingredient which would combine with common salt ions to form soluble molecules which would neither interfere with the detergent action, nor precipitate out to make a wash look tacky. Unfortunately for the lakes, this added ingredient contained phosphates, and a number of lakes are undergoing accelerated eutrophication as a result.

Synthetic Organic Compounds

Although the problem of phosphates in detergents has been mentioned, the detergent molecules themselves caused trouble earlier because their detergent action continued too long after they were flushed down the drain. Rivers, ponds, creeks, and lakes began to develop a nasty "head" of suds. A different detergent molecule, one which is broken down to simpler forms by naturally occurring processes, was substituted for the original long-lasting variety.

Another class of synthetic organic compounds, insecticides, has caused a great deal of alarm recently. DDT (its chemical name is dichlorodiphenyltrichloroethane) has caused the most alarm, so let's look at this compound briefly. Its molecule consists of carbon, hydrogen, and chlorine atoms arranged in a symmetric fashion. Because of the symmetry of the molecule, it is almost insoluble in water, but it is soluble in other symmetric molecules like oils and fats. The mechanism by which it kills insects is not completely understood, but it is thought to dissolve in the fatty tissue around nerve cells and interfere with the transmission of nerve impulses. The DDT molecule is not quickly degraded to simpler substances by natural processes. Because of this slow degradability, this compound persists long after it has killed the insects it was intended to kill. Because of DDT's insolubility in water, it sinks to the bottom and is eaten by plants or small fish, which are then eaten by larger fish, etc. The slow degradability and solubility in fat lead to a larger concentration in the fatty tissues of larger fish, until there is some potential danger to the larger fish and man. Not all "experts" agree on its effects or what to *do* about DDT, but its use is banned in some states, and alternative pesticides which degrade more rapidly are being developed and used.

Oil

The complicated mixture of organic molecules called oil has been showing up in oceans and lakes quite a lot in the last few years. Some of the molecules contained in oil *are* water soluble and slip into the water quickly after a spill. Other parts (called aromatics because they evaporate easily and present themselves to your sense of smell) evaporate into the atmosphere quite readily. Most of the oil is not soluble and spreads out over the surface of the water in a thin film. This oil coating interferes with the dissolving of oxygen needed by fish, and with amphibious creatures and waterbirds who dive into the oil attempting to catch fish. As larger and larger oil tankers are built, potentially disastrous oil spills are foreseen by some.

Inorganic Chemicals

Drainage of H_2SO_4 (sulfuric acid) from mines, particularly from abandoned coal mines, is a major source of pollution in many streams.

Certain salts are a major pollutant of freshwater lakes and rivers. Major sources of salts include industrial effluents, irrigation water, ocean water back-up, and salt used to thaw icy highways. The ocean may back up into coastal rivers during periods of low water flow and create difficulties for aquatic life, and it sometimes threatens public water supplies. We mentioned earlier the way solvated molecules get in the way and don't permit water to form a solid crystal until the temperature drops below 0° C. The salt spread on the roads to keep ice from forming in the northern states in winter has not only killed shrubbery and trees, but has polluted the wells in some rural areas.

A large amount of mercury is "lost" each year by the chemical industry in the preparation of chlorine (Cl_2) gas and by the paper industry, which uses mercury compounds to retard the growth of fungi on logs. Although mercury is toxic in nearly all its forms (the "mad hatter" of Alice in Wonderland fame may have inhaled mercury vapor since mercury was used as a dressing for felt), almost all inorganic compounds of mercury are insoluble in water. Companies thought they could discharge mercury-containing wastes into the water and they would sink to the bottom and lie harmlessly in the sludge. WRONG! Some of the bacteria that also live in the sludge convert mercury to methylmercury, which is an extremely toxic form that can be readily absorbed by plants and animals. In the 1950's, the Minamata Bay area of Japan had a serious outbreak of mercury poisoning. Fishermen, their families, and their household cats became ill after eating fish caught in the bay. A local plastics factory had been discharging mercury-containing wastes into the bay, and the concentration of mercury worked its way right up the food chain to man.

Sediments

Too much suspended material also acts as a pollutant in the sense that it tends to fill up reservoirs and channels, causing the reservoirs to overflow and flood nearby areas and causing channels to become wide but shallow—hardly ideal for navigation purposes. One of the great influences on the amount of suspended material carried away by rainwater is the amount of vegetation available to absorb the water. As modern developments continue to pave our former vegetated areas, rainwater washes into streams in larger amounts, generally taking more suspended material along for the ride.

We've seen a lot about what *could* be in the water. Let's now do the scientific bit and see for ourselves what *is* in the water.

EXPERIMENT 8.1

WHAT'S IN THE WATER BESIDES H_2O?

Various tests may be performed to analyze the contents of a given water sample, depending on the equipment and time available. It is recommended that the test equipment used be explained thoroughly to avoid the "black box" feeling on the students' part, and that samples of drinking water, pond water, river water, etc., be tested to show differences. A brief discussion of parts per million (ppm) and parts per billion (ppb) would also be helpful here.

8.4

How Can Water Become Unpolluted?

It might be very tempting at this point to think about banning this and banning that. We must be wary of any solution that is too simplistic. Let's look here at the molecular possibilities of reversing pollution and save the rest of the discussion of what to *do* for Chapter 10.

There are two fundamentally different ways to unpollute water. The first is to *prevent the pollutant from entering the water* in the first place, and the second is to permit the pollutant to enter the water, but *take it out at some later time and place.*

Industrial polluters can often use *recycling techniques* to prevent pollutant discharges. For example, the chlorine gas producers who used to discharge mercury have begun recirculating their waste water and installing lagoons and settling ponds in which the mercury collects. In 1970, their mercury discharges were reduced by almost 90%.

Urban waste-disposal systems are confronted with the problem of fully polluted water arriving at their plants in great quantity, and it must be treated and returned to the water table fairly quickly. The treatment that sewage gets is generally classed as primary, secondary, or tertiary (although about 5% of all sewage gets no treatment at all).

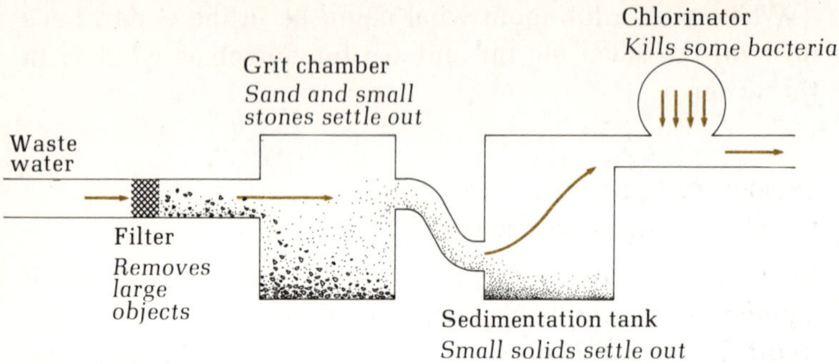

Figure 8.7 *Primary sewage treatment*

Primary treatment consists of allowing suspended particles to settle out (possibly in several stages) and then adding chlorine gas. Recall that chlorine is a very active element, needing a single electron to become a chloride ion. Added to water, it will get its electron at the expense of some of the bacteria in the water, causing their death in the process (see Figure 8.7).

Perhaps a third of the suspended solids and oxygen-demanding wastes are removed by primary treatment, but often all dissolved oxygen is used up, and anaerobic decomposition, with its attendant odors, takes place in the effluent. A *secondary treatment* is often employed to remove more of the oxygen-demanding wastes. In the *trickling filter method,* the waste water is trickled through a deep bed of stones open to the atmosphere. Bacteria living in the stones and supplied with plenty of oxygen from the atmosphere convert the oxygen-demanding wastes to inert inorganic materials. The bacteria remaining in the water are killed by chlorine, and the resulting water may have its BOD load cut by 75% (Figure 8.8).

Another type of secondary treatment is called the *activated sludge method,* and it reduces the BOD a bit more

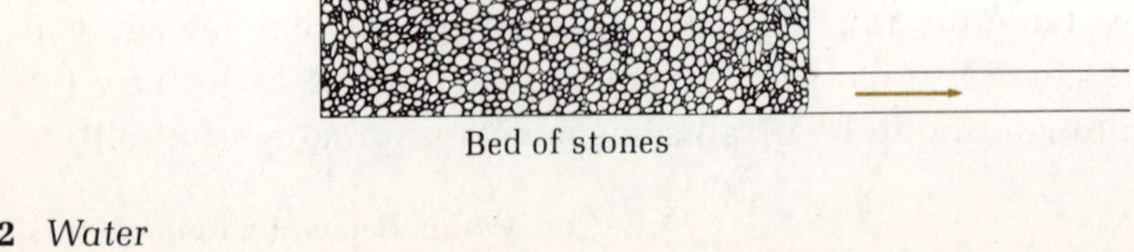

Figure 8.8 *Trickling filter secondary sewage treatment*

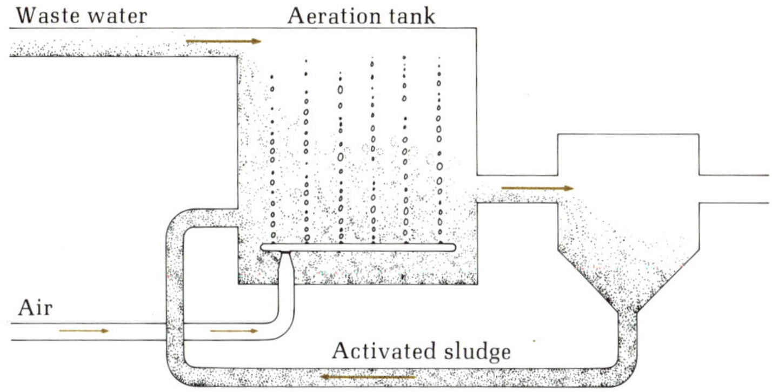

Figure 8.9 *Activated sludge secondary sewage treatment*

efficiently than the trickling filter method by bubbling air through the sewage in the presence of the bacteria already contained in the sewage (Figure 8.9).

Tertiary, or further, *treatments* may be required to remove other pollutants, but the technology is rather complex and not widely used at this time. An example is the addition of various chemicals to form insoluble precipitates with the phosphorous ions. Other examples include "activated" charcoal to take out dissolved organic compounds.

Urban water supplies are treated with various chemicals to cause suspended dirt particles to become large enough to precipitate out rapidly. The water is often filtered through sand and gravel, sometimes sprayed into the air to get more dissolved oxygen into it, sometimes passed through charcoal filters to remove organic coloring or odorous compounds, then it is chlorinated to kill bacteria.

Summary

Water has some unusual properties which make it "strange:" first, water should be a gas at ordinary temperatures. Taking the elements directly below oxygen in the periodic table, sulfur (S), selenium (Se), and tellurium (Te), these elements should have chemical properties similar to oxygen, so they should combine with two hydrogen atoms to form molecules similar to water. H_2S, H_2Se, H_2Te are all gases at normal temperatures, so one might predict that H_2O should also be a gas. The reason water isn't a gas at normal temperatures is that the water molecule is more strongly polar than the H_2S, H_2Se, or H_2Te molecules.

A second strange property is that solid water "should"

sink in liquid water. Well over 90% of all substances have a greater density in the solid state than in the liquid state. The reason ice floats has to do with the geometry of the water molecules. Because of the way the hydrogens bond onto the oxygen (at an angle just greater than 90° rather than a straight line as one might expect), the water molecule can be assembled to form ice only in a particular way as shown in Figure 8.3. The structure of ice is a very open one; the water molecules are rather far apart, resulting in a low density for ice.

A third strange property of water is that water "should" be capable of dissolving only a few substances. Most liquids dissolve only solids whose molecules are similar to the liquid's molecules in size, shape, and charge distribution. Because of the water molecule's unique shape and charge distribution, it dissolves ionic solids, substances with large natural dipole molecules, and even some substances held together by induced dipole bonds.

On the other hand, water is quite "usual" in the sense that natural waters abound in the form of rivers, lakes, and oceans. It's easy enough to see that water doesn't remain "pure" H_2O very long because of its capability for dissolving a large variety of solids. The reason that the ocean is salty is that the salt dissolved in natural water doesn't escape with the water molecules that evaporate from the ocean. The salt ion and its associated water molecule have too large a mass to escape easily, so the salt collects in the ocean. The water cycle traces the water vapor which has evaporated from the ocean in the form of clouds; as it is blown across the land, it cools down and forms liquid water which falls as rain. The rainwater drains across the land, dissolving solids as it goes until it finds a river. The river then flows, carrying the water and dissolved and suspended solids along, eventually to the ocean. Water vapor evaporates from the ocean in the form of clouds . . . (recycle).

Solid water played an interesting role in shaping the contours of northern United States in that glaciers have made at least four major advances and retreats over the last 10^6 years.

Currently, man "uses" water in a variety of ways: urban water supplies, waste disposal, energy generation, irrigation, navigation, recreation, esthetics, and various industrial uses. Unfortunately, these uses often conflict, leading to polluted water, which we have defined as water which contains substances which render the water unsuitable for an intended use.

One class of pollutants are oxygen-demanding wastes. Some organic compounds react with the oxygen dissolved in the water, depriving aquatic life of its life-sustaining oxygen. A measure of the amount of oxygen required to oxidize the organic compounds is called the biochemical oxygen demand (BOD). Some bacteria aid the decomposition of organic molecules. If oxygen is involved in the decomposition, the bacteria are called aerobic. Other bacteria, called anaerobic bacteria, help the decomposition without oxygen. As the oxygen in a body of water is depleted by oxygen-demanding wastes, anaerobic decomposition takes over, generating foul-smelling flammable gases.

Other classes of pollutants are disease-causing agents, plant nutrients which lead to accelerated eutrophication, synthetic organic compounds such as pesticides and herbicides, which do not degrade rapidly and may work their way up the food chain to man, oil, inorganic chemicals such as mercury, and sediments which clog channels and cause flooding.

The two fundamentally different ways of halting the pollution of water are either to prevent the pollutant from entering the water or to take it out at some later time and place.

Industrial polluters are sometimes able to recycle materials which they had formerly dumped into the water.

The treatment of polluted waste water, particularly by sewage treatment plants, is classed as primary, secondary, or tertiary sewage treatment.

Primary sewage treatment allows suspended particles to settle out, then chlorine gas is added to kill bacteria. Secondary treatment removes more of the oxygen-demanding wastes, and tertiary treatments are sometimes used to take out specific pollutants like phosphates.

The treatment of water for urban water supplies usually takes the form of filtration, addition of oxygen, removal of coloring or odors, and chlorination.

Questions

This may be used like a programmed text by referring to the indicated questions while reading the text.

8.1

1. For what reasons "should" water be a gas at normal temperatures?
2. Why should H_2S be similar to H_2O?
3. Why does oxygen grab hydrogen's electrons more strongly than sulfur?
4. Why "should" ice sink in liquid water?
5. For what structural reason does ice float in water?
6. Why "should" water be capable of dissolving only a few substances?

8.2

7. Why isn't natural water pure H_2O?
8. Give some examples of things carried along by natural waters.
9. What are the major differences between a river and a lake?
10. What are the major differences between a lake and an ocean?
11. What are tides and what is their cause?
12. What does the study of oceanography encompass?
13. Outline the water cycle.
14. Why don't solvated molecules evaporate along with regular liquid molecules?
15. What are glaciers?

8.3

16. List several broad categories of uses for water.
17. List some of the normal household uses for water.
18. Why is water so handy for disposing of wastes?
19. If so little electrical energy is generated by flowing water, why is flowing water considered an important source of electrical energy?
20. How is water used for navigation in modern times?
21. List several recreational and esthetic "uses" for water.
22. List some industrial uses for water.
23. Give some examples of conflicting water uses.
24. Give a working definition of *pollution*.
25. Explain what is meant by *oxygen-demanding wastes*.
26. In what sense must aquatic life compete with waste for oxygen?
27. What is biochemical oxygen demand (BOD)?
28. What is the difference between the actions of aerobic and anaerobic bacteria?
29. How does the presence of *E. coli* indicate the presence of "fresh" human or animal waste?

30. Outline the normal life cycle of aquatic plants.
31. Outline the natural aging process of lakes, called eutrophication.
32. In what way have we managed to accelerate the eutrophication process?
33. In what way did early detergents cause water pollution?
34. Why does DDT cause a problem for species other than the insects it was designed to eliminate?
35. What happens to oil spilled on the water?
36. How does salt prevent freezing?
37. Why did companies think they could discharge mercury-containing wastes into the water without causing any pollution?
38. How does mercury become converted to readily assimilated forms?
39. How does too much sediment act as a pollutant?

8.4

40. What are the two basic ways to unpollute water?
41. Give an example of a recycling unpollution technique.
42. Outline a primary waste-water treatment, indicating the things taken out of the water.
43. What is the main purpose of secondary treatment?
44. Outline the trickling filter method of secondary treatment.
45. Outline the activated sludge method of secondary treatment.
46. What would be the main purpose of tertiary treatment?
47. How are urban water supplies treated?

Quotes to Ponder

1. "No one is probably more concerned about the environment and protecting nature from pollution than the farmer because he makes his livelihood from the soil.

One farmer who has found a method of not only protecting the environment but also improving his farm operation efficiency is Ernie Girbach of 2693 Arkona Road, Saline Township.

For many years, Girbach has been coping with the problem of preventing runoff of liquid and solid animal waste into a nearby creek. Another problem was that the runoff oftentimes built up into a slightly smelly small body of water next to the house. Also, something was needed to dispose of the manure during the wet seasons.

A livestock waste disposal system was installed on the Girbach farm in August, 1971, to help eliminate these headaches.

The system consists of an eight-foot deep underground holding tank.

The tank is emptied about every three months by pumping the contents into a 3,250 gallon manure spreader. The manure is then spread on the fields when needed." (Jim Kane, *The Ann Arbor News,* October 8, 1972)

2. "Giant tankers as long as the Empire State building, 'docking' and piping off their thick black crude oil at specially built, deep-water terminals anchored one to eighty miles offshore.

It's still a vision. But reality may not be far off, as American demands for oil continue to grow. Maritime experts say a new generation of supertankers—as long as 1,400 feet and weighing 1 million tons—will handle nearly all the major world oil trade in 10 years.

No Atlantic or Gulf port is deep enough to handle such enormous vessels. So federal and local agencies are looking at controversial offshore proposals that please oil companies, but trouble some environmentalists.

The environmental concern: massive oil spills that could be more difficult to clean up in rougher offshore waters, and uncontrolled industrial development near deep-water harbors, which could result in overcrowding, land abuse, and pollution of air and water." (Christian Science Monitor Service, October, 1972)

3. "Shell Oil Company has developed a new concept in oil-skimming barges which will permit quicker and more efficient recovery of oil spills on inland waterways, it said Wednesday.

The main feature of the barge is a design called 'articulation,' which allows various barge components to flex with waves and currents. Thus, the barge is capable of moving at much greater speeds while skimming." (United Press Release, October, 1972)

4. "We are dumping vast quantities of pollutants into the oceans. According to one estimate by the United States Food and Drug Administration, these include a half-million substances; many are of recent origin, including biologically active materials such as pesticides, radioisotopes, and detergents to which the ocean's living forms have never before had to try to adapt. No more than a minute fraction of these substances or the combinations of them have been tested for toxicity to life—to the diatoms, the microscopic marine plants that produce most of the earth's oxygen, or to the bacteria and microorganisms involved in the nitrogen cycle." (La Mont C. Cole, "Can the World Be Saved?" *New York Times* Magazine, March 31, 1968)

5. "Before man, evolutionary processes of adaptation have been fast enough to assure survival of life on earth, but we can no longer look to evolution to catch up with the changes produced by man's technology. One example will give an idea of how man has accelerated terrestrial processes: In the past 50 years, Lake Erie has changed to an extent that would have required 15,000 years by the processes of natural succession. It is now considered to be in imminent danger of turning into a huge swamp." (Turk, Turk, and Wittes, *Ecology, Pollution and Environment,* W. B. Saunders Company, Philadelphia, 1972)

6. "For the first time in the history of the world, every human being is now subjected to contact with dangerous chemicals, from the moment of conception until death. In the less than two decades of their use, the synthetic pesticides have been so thoroughly distributed throughout the animate and inanimate world that they occur virtually everywhere. They have been recovered from most of the major river systems and even from streams of groundwater flowing unseen through the earth. Residues of these chemicals linger in soil to which they may have been applied a dozen years before. They have entered and lodged in the bodies of fish, birds, reptiles, and domestic and wild animals so universally that scientists carrying on animal experiments find it almost impossible to locate subjects free from such contamination. They have been found in fish in remote mountain lakes, in earthworms burrowing in soil, in the eggs of birds—and in man himself. For these chemicals are now stored in the bodies of the vast majority of human beings, regardless of age. They occur in the mother's milk, and probably in the tissues of the unborn child." (Rachel Carson, *Silent Spring,* Houghton Mifflin Company, Boston, 1962)

CHAPTER NINE

GASES AND AIR

We had to build several large concepts before we could analyze solid matter as fully as we did. Liquids were a bit easier to understand because of our experience with solids, but liquids still posed some difficulties because molecules in liquids hold together part of the time and move independently part of the time. We'll find gases a lot easier to understand because gas molecules are almost completely independent of each other. The mixing of gases with other gases, liquids, and solids will be analyzed, and we'll apply what we learn about gases to that mixture of gases we live in—*air*. We'll try to puzzle out how the earth got its atmosphere originally, how the atmosphere has evolved, what "pure" air contains, and what sort of dynamics happen to the atmosphere (weather). Last of all, we'll analyze the pollutants in our atmosphere (see Figure 9.1), how they get in, and how they can be removed.

9.1

How Do Gases Differ from Liquids and Solids?

Solids hold themselves together, maintaining a fixed shape and volume, while liquids maintain a fixed volume, but assume the shape of whatever container they are put in. Gases maintain neither a fixed volume nor a fixed shape, however. For example, if a leak occurs in the natural gas line in your home while you're gone (you know, for cooking, heating, etc.), when you get back, the whole house smells like gas. Another example involves two containers of different colored gas separated by some kind of partition that may be removed, as shown in Figure 9.2.

When the partition is removed, both gases expand into each other's container and the resulting mixture is some color in between the two original colors. Because gases, unlike solids or liquids, expand to fill whatever volume is available, we can expect some structural differences also.

Figure 9.1 *S. L. Ump predicts troubled air ahead*

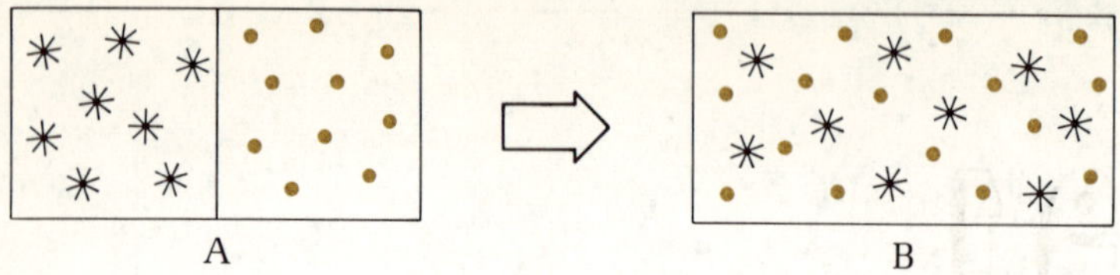

Figure 9.2 *A. Two containers of different colored gases, and B. a mixture of the two gases*

Let's ask the same questions about gases that we asked about solids and liquids. *What are gases made of?* Just like solids and liquids, gases are made of atoms and molecules. *How far apart are the atoms and molecules in gases?* To answer this one, we'll have to take advantage of your powers of observation. Picture a pot or a kettle of water boiling vigorously. The process of boiling is the conversion of a liquid into a gas, so gaseous water (steam) is formed. The steam forms as bubbles at various points in the pot or kettle, and these bubbles rise to the surface of the water. Since the steam bubbles rise, they must have a lower density than liquid water. Since density is mass/volume, steam must have less mass in a unit volume than liquid water. Less mass means fewer water molecules, so a gas must have a *larger spacing between atoms and molecules than either liquids or solids.* Furthermore, energy must be supplied to a liquid to change it into a gas, and this energy shows up as increased potential energy in the gas. Recall that increased potential energy means increased distance between molecules. *What holds gases together?* In solids and liquids, electrical forces between molecules kept the molecules occupying points in some crystal structure or kept the molecules maintaining some relationship with other molecules; although in both cases, the molecules could move around somewhat as indicated by their temperature and hence kinetic energy. Yet gases don't maintain any fixed volume or shape, so their molecules really aren't being held together in the same sense as a liquid's or solid's molecules. The molecules in a gas are much more independent than the molecules in a liquid or solid.

Let's use the following as a conceptual model for a gas:

A gas consists of a collection of molecules in constant, random motion.

First let's look at some of the implications of this model in terms of what Newton's laws tell us about moving bodies.

Let's focus our attention on a single molecule of a gas

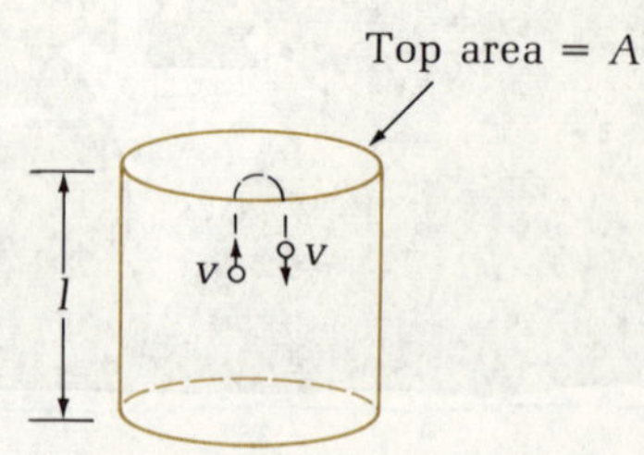

Figure 9.3 *Single gas molecule before and after collision with the top of a cylinder*

inside an enclosed cylinder. Suppose the molecule moves upward at a velocity, v, and "collides with" the atoms or molecules of the container as shown in Figure 9.3. (Although we know that what really happens is that electron clouds overlap until a repulsion is built up, it is convenient to speak of a "collision" because it is a simpler concept to envision.) If we assume that no kinetic energy is lost, then the collision is said to be *elastic,* and the molecule moves back downward with the same velocity, v, so that the kinetic energy after the collision, $\frac{1}{2}mv^2$, is the same as the kinetic energy before the collision, $\frac{1}{2}mv^2$.

When the molecule collides with the top of the cylinder, the cylinder must experience a force. On the other hand, the molecule feels an acceleration as it is turned around from going upward to going downward. The relationship between the force exerted by the cylinder top on the molecule and the molecule's resulting acceleration is given by Newton's second law:

$$F = ma.$$

But a is simply the change in velocity divided by the time it took to change (recall Chapter 4), and the change in velocity is from v up to v down, a total of $2v$ (think about it!).

$$F = m\frac{2v}{\Delta t},$$

where Δt is the time over which the collision occurs.

This force is a peak force and only occurs when the molecule is colliding with the top of the cylinder. Let's look at an average force over the time interval that it takes to get from the top of the cylinder to the bottom and back to the top. The total distance traveled $= 2l$; the velocity of travel $= v$; the time of travel is $2l/v$.

$$F_{\text{average}} = m\frac{2v}{2l/v} = \frac{mv^2}{l}.$$

Now, let's divide the average force by the area of the top of the cylinder, A. Force/area is called *pressure, p.*

$$\frac{F_{\text{average}}}{A} = \text{pressure} = p = \frac{mv^2}{lA}.$$

lA is simply the volume of the cylinder, V; so the equation becomes

$$p = \frac{mv^2}{V} \quad \text{or} \quad pV = mv^2.$$

So far, we've been concerned with a single molecule moving up and down. If we take into account the facts that we have a large number of molecules moving in all directions, and that the total kinetic energy is directly proportional to the temperature, we may arrive at the relationship

$$pV \propto T.$$

This says that for the idealized gas we are considering, the pressure exerted by a gas multiplied by its volume is proportional to its kinetic energy or temperature. (The mathematical development of this equation was included to show you that there is nothing magical about the final equation $pV \propto T$. It came directly from Newton's second law and our assumption about the kinetic energy being conserved in collisions.) This *prediction* about the pressure, volume, and temperature of an ideal gas has been tested in a variety of classical *experiments*. Let's do one.

EXPERIMENT 9.1

FOR A CONSTANT VOLUME OF GAS, IS PRESSURE DIRECTLY PROPORTIONAL TO TEMPERATURE?

The apparatus required for this experiment consists of a metal sphere with a pressure gauge and an insulated handle (Figure 9.4).

A gas is sealed inside the metal sphere, and the pressure of the gas is recorded by the pressure gauge. (It shouldn't matter what gas it is, because our ideal gas model doesn't make any distinction between one gas and another.) Now, we could measure the pressure by the gauge and plot the pressure vs. the temperature to see if they are proportional. To cause this temperature variation, let's use our old friend, water. If we start with a beaker of ice water and a thermometer, we can immerse the metal sphere containing the gas, and it will eventually reach the same temperature as the water. If we heat the water slowly (heating the water too rapidly will cause the water temperature to be higher than the temperature of the gas inside the metal sphere), we can record the temperature of the water (and hence the gas temperature also) and the pressure of the gas as the water is heated from its original 0°C to its boiling point of 100°C. Take temperature and pressure readings at one-minute intervals until 100°C is reached. Plot the results on a graph arranged as shown in Figure 9.5.

Draw the best single straight line through the points,

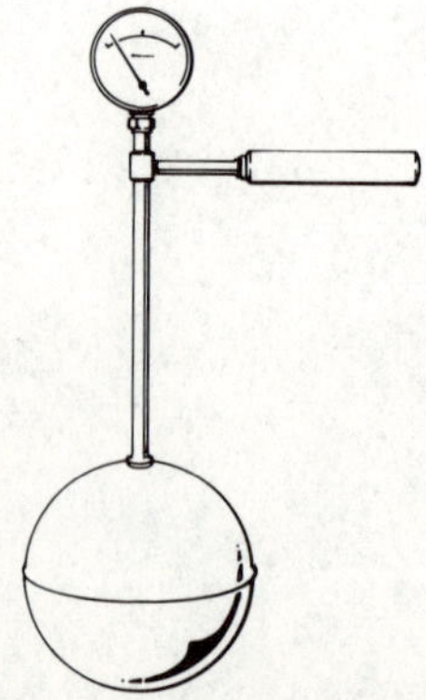

Figure 9.4 *Apparatus for Experiment 9.1*

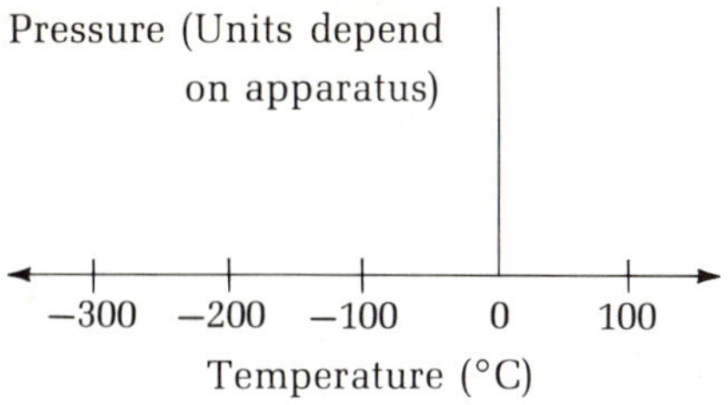

Figure 9.5 *Graphical format for Experiment 9.1 data*

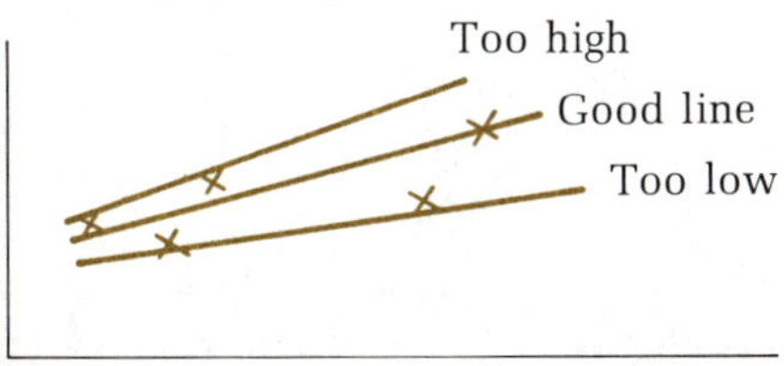

Figure 9.6 *Example of a single straight line through a series of data points*

attempting to minimize any differences between the straight line and the points. (See Figure 9.6.)

Once your straight line is drawn, extrapolate the line by simply continuing it below 0°C all the way down to zero pressure (Figure 9.7).

Extremely careful experiments performed by others have shown that pressure is proportional to temperature, and that the temperature at which the pressure goes to zero is about −273°C. This is called absolute zero, and since the temperature and kinetic energy are proportional, the kinetic energy and hence velocity also become zero. At absolute zero, molecular motion would cease. Although there are reasons (based on the Heisenberg uncertainty principle) for believing that absolute zero cannot be reached, it has been approached in many laboratories. As you might well imagine, some very unusual things happen at extremely low temperatures, with solids exhibiting unusual magnetic properties, liquids flowing uphill, and normally pliable solids becoming brittle enough to shatter. The physics of low temperatures is called cryogenics.

Another phenomenon which tends to confirm the notion that gas molecules are in random motion is *Brownian motion.* Did you ever notice an extremely small dust particle in strong sunlight? It bounces around rather than moving in straight lines. Although you can't see what's bouncing it around, it is the gas molecules of the air moving in random motion.

How about the potential and kinetic energies in gases? Recall that the depth and spacing of potential energy wells was the chief difference between the various kinds of solid matter. In the liquid state, the potential wells were all irregular, and only a portion of the molecules were ever in the

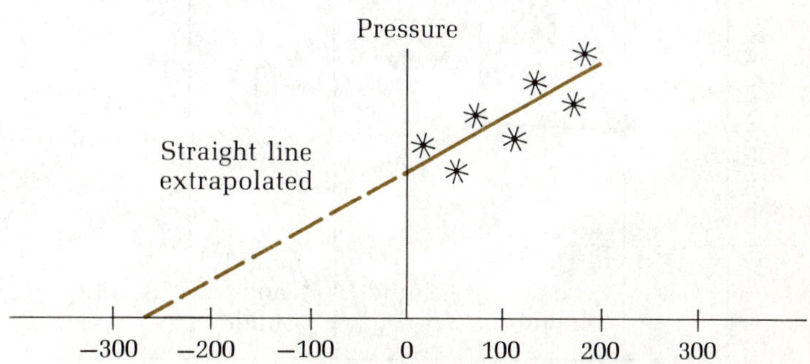

Figure 9.7 *Extrapolation to zero pressure*

potential wells at any one time, while the remainder of the molecules were free to move about. The chief difference between a liquid and a gas, in terms of energy, is that *no* molecules are trapped in any potential wells in a gas. To go from a liquid to a gas, enough energy must be supplied to the liquid to lift all the molecules out of potential wells so that they can become free to move in any direction, and hence become part of a gas. There was a *melting point* at which a solid remained at a single temperature while enough potential energy was supplied to tear down the solid structure and move the molecules farther apart to form a liquid. Similarly, we can expect that there will be a *boiling point* at which the liquid will remain at a single temperature while enough potential energy is supplied to tear apart the liquid structure and move the molecules even farther apart to form a gas. An easy way to test this would have been to measure the temperature of the boiling water during Experiment 9.1. Of course, it should be 100°C. That's why cooking with water gives reasonably reliable results—the temperature involved is constant.

We are finally in a position to look at the potential and kinetic energy structures for all three phases of matter: solid, liquid, and gas. Figure 9.8 summarizes what we have learned.

The ideal gas model we've been working with is really a pretty good one. Its flaws show up only in very dense gases where some of the collisions become inelastic, causing the atoms contained in the molecules to rotate and vibrate rather than sending the molecule as a whole off in some random direction. When the temperature gets too high, the model also breaks down because the collisions get violent

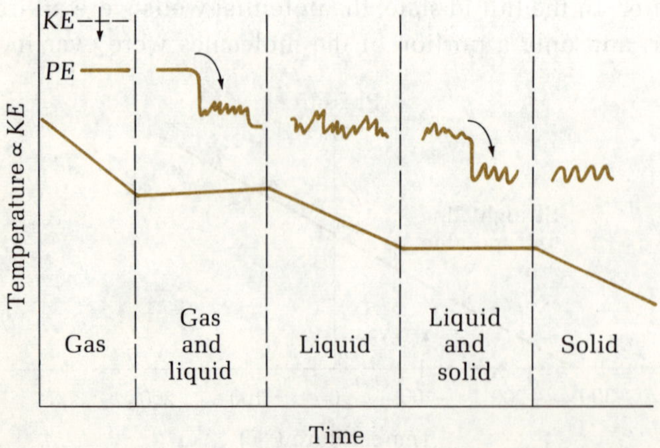

Figure 9.8 *Potential and kinetic energy diagrams for a substance as it cools from a gas to a liquid to a solid*

enough to start ripping some molecules apart, and these torn-up molecules may combine with others, altering the molecular structure of the gas.

9.2

How Do Gases Get Together with Solids, Liquids, and Other Gases?

Since gas molecules are quite independent of each other, exerting forces only through collisions, we would expect gases to deal with solids, liquids, and gases in pretty much the same way. Solids and liquids immersed in gases find their surfaces bombarded by gas molecules in constant, random motion. If the solid or liquid object wasn't there, a "chunk" of gas the same shape and volume as the object would have been there, as shown in Figure 9.9. The "chunk" of gas would have stayed right there, so the total force exerted on that "chunk" by the surrounding gas must be equal to the weight of the "chunk."

The presence of the object doesn't change the gas pressure; the pressure is the same as if the gas chunk were still there. Therefore, the upward force exerted by the gas is still the same. If this upward force is greater than the weight, then the object is said to be *buoyant*. If we examine this force difference mathematically, we find:

Net upward force = force due to gas pressure − weight of object
= weight of chunk of gas − weight of object
= (mass of chunk of gas)g − (mass of object)g
= (density of gas)(volume of chunk)g − (density of object)(volume of object)g.

But the volume of the chunk and the volume of the solid are equal, so

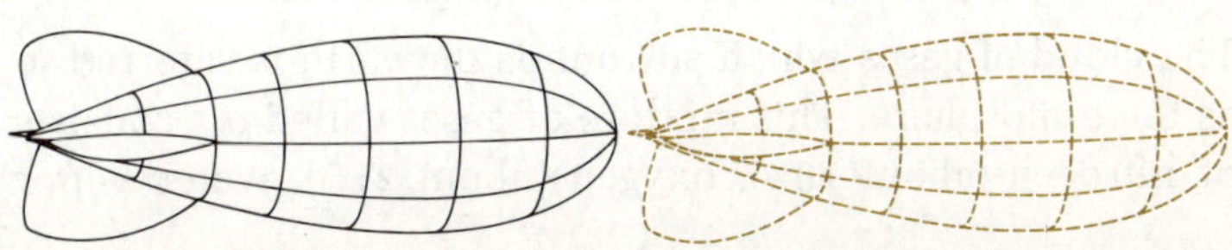

Figure 9.9 *Object immersed in gas and "chunk" of gas same shape and volume as object*

Net upward force

$$= (\text{density of gas} - \text{density of object})(\text{volume})(g).$$

Just as we suspected, for an object to float, it must have a smaller density than the medium in which it wishes to float. (The same line of reasoning holds for liquids as well as gases.)

Solid and liquid objects with a density larger than the gas in which they are located will, therefore, sink, but their sinking may be influenced by two major factors.

1. *Overall motion of the gas:* If an object is sinking with respect to the gas, but the gas is rising with respect to the earth, the object's rate of sinking may be very low, or it may even be rising with respect to the earth. Sailplane pilots depend on finding air rising faster than the sinking speed of their plane.

2. *Friction between the object and the gas molecules:* As an object sinks faster and faster, it collides with more and more air particles and loses some of its kinetic energy to each one. Eventually, a balance is reached where the potential energy lost by the object as it falls doesn't show up as increased kinetic energy of the object, but is passed along to the gas and turns up as increased kinetic energy of the gas. The maximum speed attained by an object, called *terminal speed,* is thus determined by the interaction between the falling object and the gas it's falling through, not by the height of the fall. (That's why raindrops from high clouds don't hit your head any harder than raindrops from low clouds.)

Since gases are so independent at the molecular level, you'd expect *gases to mix with each other in practically all proportions* until the density or the temperature gets too high and the molecules start breaking apart.

9.3

That Mixture of Gases We Live in—Air

The cloud of gases which surrounds the earth is referred to as the *atmosphere*. This mixture of gases called *air* consists of nitrogen (about 78%), oxygen (about 21%), water vapor

(0–3%), and less than 1% of argon, carbon dioxide, neon, helium, methane, krypton, hydrogen, and others. Hundreds of kilometers thick, this blanket of atmospheric gases performs a number of functions which are critical to life on earth.

ANXIOUS STUDENT: But, Professor, where did air come from originally?

PROFESSOR: The answer to that one isn't entirely clear. Recall the model proposed in Chapter 6 for the formation of the earth. Atoms formed from the "Big Bang" and from the remnants of exploding stars (supernovas) came together because of gravitational attraction to form the earth. Because the atoms came closer together, they lost potential energy and gained kinetic energy, so the earth had a high temperature at that time. The lower density material which floated to the surface of the earth didn't conduct the kinetic energy very rapidly; thus the earth's surface cooled down and some solids were formed. When the solid surface cooled down, liquids were formed from the lighter gaseous material around the fringes of the forming earth.

STUDENT: You mean that stuff was always 78% nitrogen and 21% oxygen?

PROFESSOR: Oh, *no!* It was undoubtedly very different from the atmosphere as we now know it. We can only speculate on the role of volcanoes, chemical reactions between atmospheric constituents and liquids or solids on the earth's surface, radioactive decay of atoms, the influence of plants, bacteria, and other forms of life, and the influence of radiation from the sun. All these things have acted over the years to finally produce the atmosphere we know today.

SUSPICIOUS STUDENT: If hydrogen and helium were present in such large quantities in the initial fireball as the universe was formed, why is there so little hydrogen and helium in our atmosphere now?

OLD PROFESSOR: Remember that the kinetic energy of the molecule is $\frac{1}{2} mv^2$ and is proportional to the temperature. In the early stages of the earth's formation, the temperature of the atmosphere was much higher than it is now, so the molecules had lots of kinetic energy. Consider the difference between a molecule with large mass and a molecule with small mass, both having the same kinetic energy. Which molecule would have the greater velocity? The light molecule would have to go faster to achieve the same kinetic energy. If the light molecule goes fast enough in the right direction, it can escape the earth's gravitational pull and become free. Massive molecules move more slowly, so fewer massive molecules escape. Hydrogen and helium are the lightest gases of all, so they escape most easily. (This is why some of the smaller planets are thought to have almost no atmosphere, and some of the larger planets are thought to have many light gases.)

SHARP STUDENT: Do you mean that the gases our atmosphere contains are just ones that are too massive to escape?

OLD PROFESSOR: Not exactly. There are some clearly identifiable processes that involve the gases in the atmosphere. For example, as illustrated in Figure 9.10, carbon dioxide (CO_2) is a natural result of the life processes of man and animals and is also given off by internal combustion engines. Green plants require CO_2 and give off oxygen (O_2) as a by-product. Man, animals, and internal combustion engines require O_2 and give off CO_2 as a natural result of the life processes . . . . (Start six lines back and repeat the cycle as often as you like.)

SHARP STUDENT: The atmosphere isn't really uniform, is it? I seem to recall hearing something about different layers which do different things.

PROFESSOR: No, the atmosphere isn't really uniform in composition, pressure, temperature, etc. Let's start out by seeing what would happen to the pressure and temperature if the composition were uniform.

The pressure of an air column is simply its weight divided by its area, so we would expect the air pressure to be lower in the mountains because there is a shorter air column above them, as shown in Figure 9.11. Thus, we would expect the air pressure to decrease with increasing altitude above the earth's surface. Furthermore, considering air to be a perfect gas, the pressure and temperature should decrease together because they are proportional to each other. However, the temperature doesn't just decrease.

SUSPICIOUS STUDENT: Wait a minute . . . I mean, excuse me, but I've been up in airplanes and I've seen the late movies with the bomber pilots all bundled up against the cold air up high in the atmosphere. How can you say the temperature doesn't just decrease?

PROFESSOR: Well, it does decrease for a while, but then it

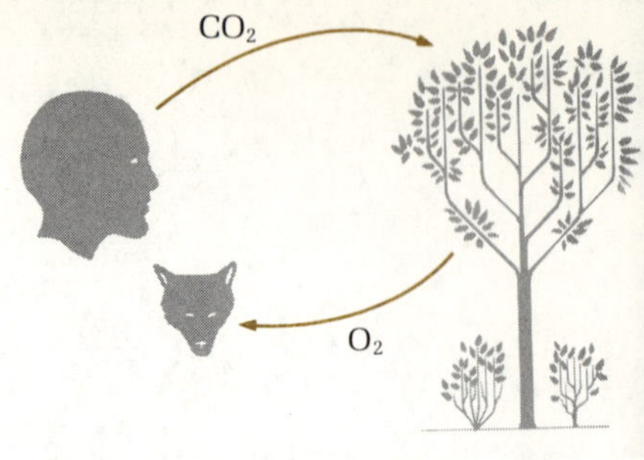

Figure 9.10 *O_2 and CO_2 cycled between animals and plants, using the atmosphere as a "carrier"*

Figure 9.11 *Air columns above lowlands and mountains*

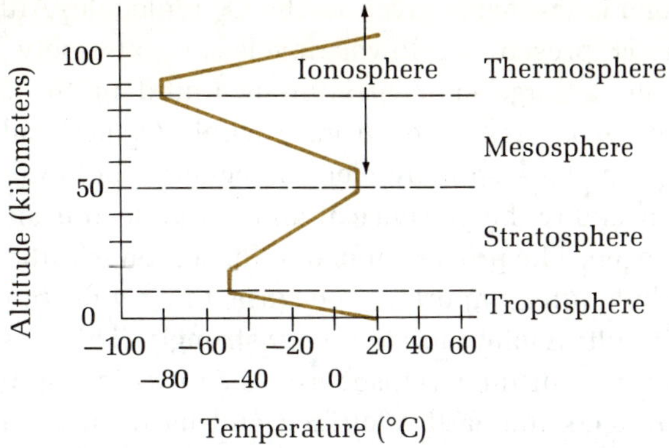

Figure 9.12 *Variation of atmospheric temperature with altitude*

levels off, and then it increases . . . oh well, here's a graph of measured atmospheric temperature to show you what I mean.

As you see from this graph in Figure 9.12, the air doesn't even begin to warm up until an altitude of 20 kilometers is reached, and the cruising altitude of the big jets is around 10 kilometers.

STUDENT: Those are some pretty wide swings in temperature in that graph. Where did we go wrong expecting a steady temperature decrease?

PROFESSOR: We neglected two very important items—*the sun,* and *ozone.* The sun gives off electromagnetic energy and gives plenty of kinetic energy to the few molecules in the highest altitudes, above 100 kilometers or so. The molecules in this region are generally torn apart by the sun's high-energy photons, and they separate into ions and free electrons. This region is called the *ionosphere,* and it has a significant effect on radio broadcasting. Recall from Chapter 3 that ions in crystalline solid matter reflect visible light very smoothly because the visible light's wavelength is larger than the spacing between ions. In the ionosphere, the ions are quite far apart, so they don't reflect light waves. However, longer wavelength electromagnetic waves, long radio waves, *are* reflected by the ionosphere (TV and FM have too short a wavelength to be reflected), allowing radio transmissions between continents. Because the ionosphere is affected so strongly by the sun, sunspot activity is reflected in the behavior of the ionosphere and hence in long-distance radio communications. (Communications from satellites have taken over some of the communications relay functions served by the ionosphere.)

Ozone is the name given to the O_3 molecule. Although ozone is present at lower levels in extremely small amounts, a large concentration is found in the *stratosphere*. Ozone is formed from a whole O_2 molecule and an O_2 molecule split by electromagnetic radiation from the sun, and in the presence of another molecule, an oxide of nitrogen. The process of ozone formation is rather delicate, but quite important, because the ozone molecule absorbs ultraviolet radiation very strongly. This raises the temperature of the stratosphere, as shown in Figure 9.11, and protects the earth's surface and us from ultraviolet radiation. High-velocity winds have been discovered in the stratosphere, but their origin is not well understood.

SHARP STUDENT: The lowest layer of the atmosphere should be fairly well understood. It seems like various measurements are being made constantly from weather stations, airplanes, balloons, rockets, and now even by satellites.

OLD PROFESSOR: We do measure the heck out of it, but our understanding of the information sometimes leaves a lot to be desired.

Wind and water vapor play the major roles in the activity of the air in the atmospheric layer closest to the earth, called the *troposphere*.

The wind arises mostly from the simple fact that warm air rises. (Ask any politician.) For a perfect gas at constant pressure, temperature and volume are directly proportional. So, when the temperature goes up, the volume goes up. Since density is mass/volume, when the volume increases, the density decreases. The sun heats the equator directly and the poles only slightly, thus one would expect warm air to rise from the equator and sink at the poles, leading to the circulation pattern shown in Figure 9.13.

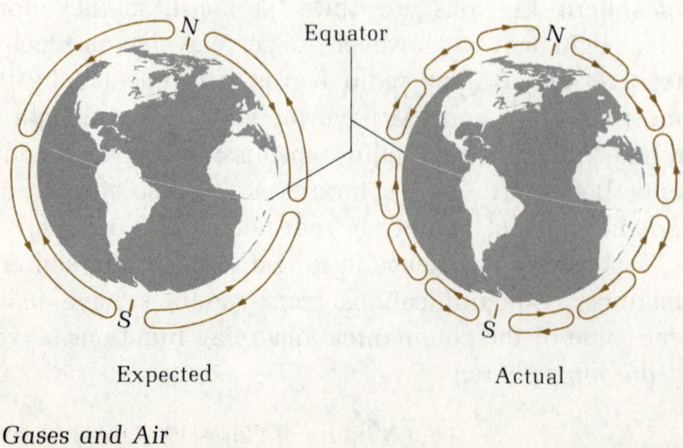

Figure 9.13 *Expected and actual air circulation patterns*

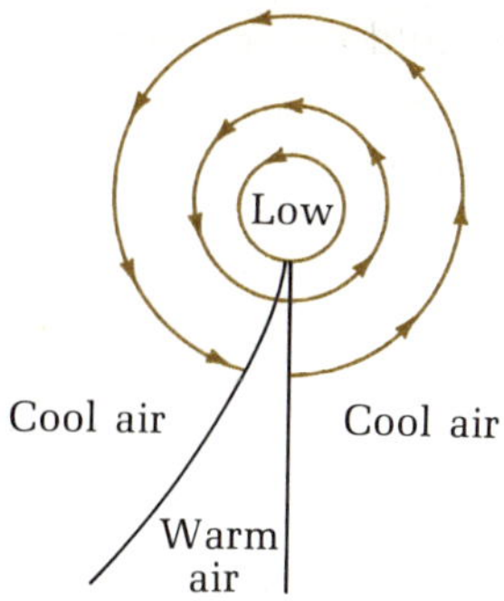

Figure 9.14 *Northern Hemisphere low (air circulates the opposite way in Southern Hemisphere)*

The actual air-circulation pattern is even a bit more complicated than shown in Figure 9.13 for reasons that are probably related to the earth's rotation, but they are not well understood. Other results of the earth's rotation and the air-circulation patterns are prevailing easterly winds in the vicinity of the equator and westerly winds in the temperate zones. Besides the prevailing winds, which affect the overall climate of a region, large air masses move under the influence of topographical features such as mountains, deserts, bodies of water, etc., which absorb and reradiate the sun's electromagnetic energy differently.

Since air is influenced by the earth below and the sun above, winds take air masses of different temperatures and pressures and run them together, often forming regions of low pressure (called *lows* on weather maps) in which winds circulate (counterclockwise, in the Northern Hemisphere) moving air masses of different temperatures into each other, as shown in Figure 9.14.

By contrast, a high-pressure zone circulates air in a clockwise direction in the Northern Hemisphere. If warm-air and cold-air masses moving together were all that weather forecasters had to worry about, their life would be very simple indeed. Our old friend water comes into this picture in the form of water vapor, clouds, fog, dew, rain, snow, hail, and sleet.

Remember the evaporation explanation from Chapter 7? High kinetic energy water molecules escape from liquid water and mix with the air molecules above. There is a limit to the number of water molecules that air will accommodate, however. This amount, called *saturation,* increases with increasing temperature. Thus, warm air may contain much more water vapor than cold air. The relationship between the amount of water vapor a given sample of air holds, and the amount it could hold if it were saturated is called *relative humidity* and is expressed in terms of a percent (100% equals saturated, 0% equals no water vapor). Let's examine the effect humidity has on the interaction between warm air and cold air masses, as shown in Figure 9.15.

As the cold air pushes into and under warm air, the warm air rises, moving to a region of lower temperature. The air which has risen can contain less water vapor because of its lowered temperature, so the additional moisture above the saturation content forms a cloud. Similarly, when a warm-air mass pushes into and over cold air, the warm air rises, lowering its temperature and hence its ability to hold water.

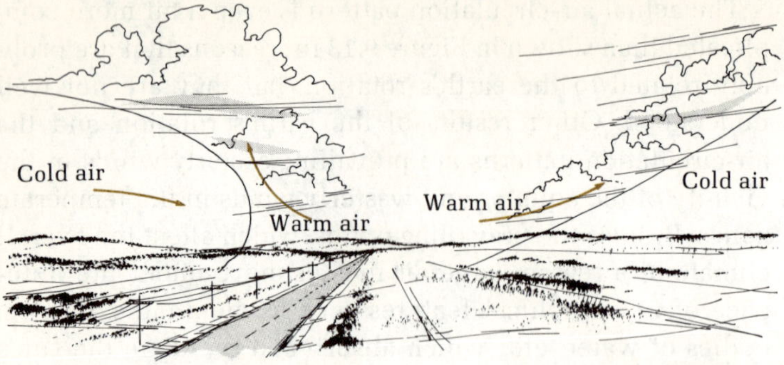

Figure 9.15 *Cold air moving into a warm air mass and warm air moving into a cold air mass*

Different kinds of clouds are formed from the excess water vapor. Depending on the amounts of water vapor involved, the height of the clouds, the temperatures of the air masses, the presence of particles on which the water could collect, etc., the water may remain as a vapor (clouds or fog), or it might form a liquid and fall (rain or sleet), or it might form a solid and fall (hail or snow). Falling liquid or solid water is called *precipitation,* and it often happens at just the wrong time, as shown in Figure 9.16.

9.4 What Pollutants Are Going into the Air, and Where Are They Coming from?

Now that we understand some things about gases and the normal constituents of the atmosphere, we can look at some of the *pollutants* (either *abnormal things or abnormal amounts*) present in our atmosphere and their sources. There are five primary pollutants: carbon monoxide (CO); hydrocarbons; the oxides of sulfur (SO_2 and SO_3, but they will be abbreviated SO_x); the oxides of nitrogen (NO_2 and NO, abbreviated NO_x); and particulate matter, in order of increasing toxicity.

Carbon Monoxide (CO): Carbon monoxide is a colorless, odorless gas at normal temperatures. It generally occurs in combustion processes in which insufficient oxygen is present to completely oxidize carbon to carbon dioxide (CO_2). The exhaust from motor vehicles is the largest single source

Figure 9.16 *S. T. Ump, weather forecaster*

of CO, generating more than double the amount put out by industrial processes and open agricultural burning. CO's toxicity to humans is related to its chemical reaction with hemoglobin in the blood. Hemoglobin normally carries oxygen, but has a greater affinity for CO, which ruins its oxygen-carrying role.

Some microorganisms present in soil absorb CO, but most of the CO is generated in urban areas rather than where there are large open soil areas for CO absorption. Reduction of CO (and other pollutants to be discussed) emission from automobile exhausts is being pursued from four fundamentally different directions: (1) modify present internal combustion engines to reduce pollutant formation; (2) develop exhaust "reactors" which will complete the combustion process and remove pollutants; (3) develop low-pollutant fuels; (4) develop an energy source other than the internal combustion engine.

Hydrocarbons: As the name implies, molecules made of carbon and hydrogen constitute pollutants in our atmosphere. Because of carbon's ability to form single, double, and even triple bonds by sharing electrons, a large number of molecules may be formed from carbon and hydrogen. Many of these molecules are gases or liquids at ordinary temperatures and escape directly, or evaporate, into the atmosphere. Of the large variety of hydrocarbons in this category, their effects on living things range from causing cancer, through noxious or mildly irritating odors, to being harmless. About half the hydrocarbon emissions come from automobiles, particularly from unburned fuel and evaporating gasoline from carburetors, gas tanks, service station pumps, etc. Better combustion techniques and venting techniques are being pursued to cut down emissions.

Sulfur Oxides (SO_x): SO_2 and SO_3 are both colorless gases at normal temperatures. SO_2 has a distinctive odor and SO_3 is extremely reactive, combining with water vapor to form sulfuric acid. Nearly 75% of the SO_2 comes from burning coal and oil. Sulfur dioxide impairs breathing in large enough doses and constitutes a serious hazard to the elderly or to those with chronic respiratory diseases. Measures being taken to reduce SO_x emissions include: (1) using fuels with low sulfur content; (2) substituting other energy sources for fuel combustion; (3) removing sulfur from fuels before burning; (4) removing SO_x from flues after burning.

Nitrogen Oxides (NO_x): NO, a colorless, odorless gas, is present in the atmosphere in far larger quantities than NO_2, a reddish-brown pungent-smelling gas. Since nitrogen and oxygen are both normal components of air, they can come together to form NO, but this can be done only at high temperatures. Since the combustion process takes place at high temperatures, almost 90% of the NO results from automobiles and stationary fuel combustion sources. NO_2 is much more toxic than NO, but the greatest hazard posed by NO_x has to do with the reactions that take place between NO_x, some hydrocarbons, and atmospheric oxygen in the presence of sunlight. *Photochemical oxidants* are formed, and these are among the most harmful ingredients in *smog*. NO_x emission control is not well in hand yet, but some steps are being taken with catalytic muffler systems for autos and systems involving water absorption of NO_x for stationary fuel-burning sources.

Particulates: Small chunks of solids and droplets of liquids with a wide range of chemical and physical properties are lumped together under the single term, particulates. As we noted earlier, solids and liquids tend to sink in gases like air, so we can generally expect the larger hunks to settle out of dust and asbestos particles from construction, fly ash from smelting and mining operations, and smoke from incomplete combustion processes. The smaller hunks don't settle nearly as readily, and atmospheric motion in the form of wind may keep some particulates up a long time. Particulates may be harmful intrinsically, like pollen to a hay-fever sufferer, or they may provide a base for another harmful substance to float around the air, either combined chemically with the particulate, dissolved within it, or sitting on its surface. Since smoke is one of the largest offenders, much effort has been directed at removing the visible particulates from smoke, often using an *electrostatic precipitator*. Since particulates are often charged, or may consist of dipole molecules, they can be attracted to a charged plate and collected. Another way to get the particulates out is to swirl the air around too rapidly for the more massive particulates to follow, causing them to bang into the sides of the swirl chamber.

Another effect of particulates is their reflection of the sun's rays, reducing the amount of the sun's energy reaching the earth. Some predict another ice age if too many particulates shield us from the sun's energy. Working in the opposite direction is the increasing amount of CO_2 in the atmosphere. The sun's rays go right through CO_2, but the longer wavelength radiation given off by the earth is absorbed by the CO_2, heating up the atmosphere. This is called the *greenhouse effect,* because CO_2 acts in the same way as the glass in a greenhouse, admitting the sun's rays but reflecting heat rays back inside, as shown in Figure 9.17.

Although particulates and CO_2 are affecting the earth's overall heat balance in opposite directions, it's not at all clear that one will "win" or that they might cancel each other out. Since particulates are more noticeable, one might almost suspect there will be more effort to eliminate them than the invisible, harmless CO_2.

Figure 9.17 *Greenhouse effect*

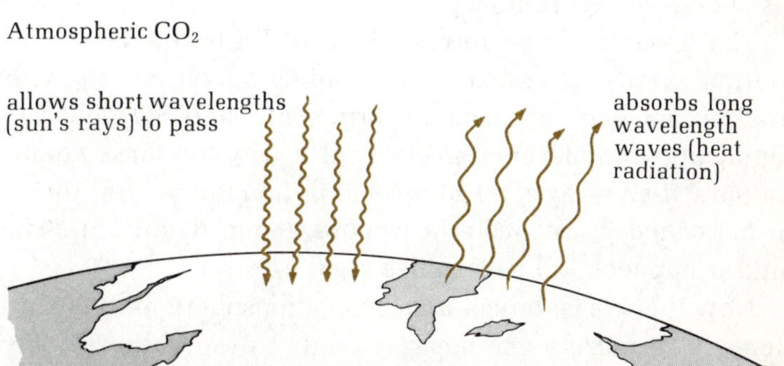

Figure 9.18 *Temperature inversion layer*

The events which have really called our attention to air pollution over the last few years have to do with a weather phenomenon called an *inversion layer*. Because of topographic features, some places often experience a warm air mass which slides completely over a cooler air mass. The temperature rises in this layer rather than falling as it normally does, and the air molecules don't circulate upward as they normally would, moving from regions of high molecular activity (kinetic energy) toward regions of low activity, as shown in Figure 9.18.

The inversion layer forms a kind of lid on the area, and normal winds are stifled. Smoke builds up, NO reacts with oxygen and hydrocarbons to form smog, and SO_2 and CO build up to intolerable levels until a new air mass comes to blow it all away. This happened in London several times, it happened disastrously in Donora, Pennsylvania in 1948, and it happens in Los Angeles regularly.

Now that we've talked about the atmosphere and its contents, let's make some measurements to see if there's any truth to what we've been saying.

EXPERIMENT 9.2
AIR SAMPLING

Various air-sampling techniques and equipment may be used depending on the time and money available. It is recommended that the techniques used be explained thoroughly and that air samples from various places, at various times of day, be analyzed to show differences.

Summary

Gases consist of molecules, farther apart than the molecules in a liquid or solid, as evidenced by gas bubbles rising in a liquid. The molecules don't seem to be held together because they take the shape of whatever container they are in, and they expand their volume to fill the container. Thus, our conceptual model of a gas is a very simple one: it is a collection of molecules in constant, random motion.

A consequence of this conceptual model plus the additional assumption that collisions of gas molecules conserve kinetic energy yields the relationship $pV \propto T$, where p is the pressure of the gas, V is its volume, and T is its temperature. This relation was confirmed by Experiment 9.1. Additional confirmation of the notion that gas molecules are in random motion is the phenomenon of Brownian motion, in which the bouncing motion of a small particle is attributed to collisions with molecules.

Potential energy must be supplied to lift molecules out of liquid potential energy wells to make a gas, so we can expect another "flat spot" in the heating or cooling curve for a substance while its molecules change their potential energy.

If a solid or a liquid is immersed in a gas, the gas molecules just bombard the surface in the same way that they would if the object wasn't there. Since a chunk of gas of the same shape as the object would be supported by the gas, the net effect of the collisions of the gas molecules with the object is to exert an upward force equal to the weight of the gas displaced. If the object's weight is less than the displaced gas weight, the object floats. If the object is heavier than the weight of the displaced gas, it will sink, with its sinking rate influenced by overall motion of the gas molecules or by friction between the object and the gas molecules.

The air surrounding the earth, called the atmosphere, is now composed primarily of nitrogen and oxygen. There was undoubtedly some kind of atmosphere present in the early stages of the earth's formation, and this early atmosphere was undoubtedly influenced by volcanoes, chemical reactions, radioactivity, plants, animals, and radiation from the sun. Hydrogen and helium from the initial "Big Bang" are not present to any great extent in the earth's atmosphere because they have small masses and can thus escape from the earth's gravitational pull rather easily.

CO_2 and O_2 are cycled between plants and animals using the atmosphere as a carrier.

Since atmospheric pressure is simply the weight of an air column divided by its area, the pressure is lower at high altitudes because the air column is shorter. The temperature of the air initially decreases with increasing altitude, then the temperature varies depending on the composition of the various layers because of radiation from the sun and the presence of ozone. Wind and water vapor play major roles in the activity of the atmosphere in its lowest layer.

Air pollutants will be considered to be either abnormal things or normal things present in abnormal amounts in the atmosphere. Five primary pollutants will be considered: carbon monoxide, hydrocarbons, the oxides of sulfur, the oxides of nitrogen, and particulate matter.

Carbon monoxide is generated mostly by automobiles and reduces the capability of human blood to carry oxygen. Carbon monoxide emission levels are being lowered by modification of present auto engines, development of exhaust reactors, use of low pollutant fuels, and the development of alternate energy sources.

Hydrocarbons are also emitted by automobiles, and their effects range from mildly irritating to cancer-causing. Better combustion and ventilation may reduce hydrocarbon emissions.

Sulfur oxides come from coal and oil combustion, and can cause difficulty in breathing. Low sulfur content fuels and alternate energy sources are being explored to reduce SO_x emissions.

Nitrogen oxides participate in the formation of smog, and reduction of their emission is not well in hand either from automobiles or from stationary combustion sources.

Particulate matter, primarily in the form of smoke, may be intrinsically harmful, or it may provide a base for other harmful substances. Because smoke particles are often charged or may have a charge induced, an electrostatic precipitator is sometimes used to reduce particulate emission. Because particles reflect the sun's rays, some say that less of the sun's energy will reach the Earth, leading to another ice age. On the other hand, increased amounts of CO_2 heat up the earth's atmosphere, creating what is known as the greenhouse effect.

The phenomenon which has called most attention to air pollution in recent years is the inversion layer, in which a region of warmer air blankets a cooler air region due to topography, prevailing winds, etc. This leads to a lack of normal circulation; and the atmosphere's dissipative mechanisms allow the pollutants discussed above to build up disastrously.

An air-sampling experiment was undertaken to find out what is in the air.

Questions

This may be used like a programmed text by referring to the indicated questions while reading the text.

9.1

1. In terms of the observable properties of volume and shape, how does a gas differ from a liquid and a solid?
2. Use observations and reasoning to show that the molecules in a gas must be farther apart than the molecules in a liquid or solid.
3. Describe our first approximation to a conceptual model for a gas.
4. What is pressure?
5. What does it mean to *extrapolate* a line on a graph?
6. At what pressure does the absolute zero of temperature occur?
7. What is Brownian motion, and how does it relate to our model of a gas?
8. In what way are melting points and boiling points similar?
9. What are the flaws in our ideal gas model?

9.2

10. Explain why the force exerted on any "chunk" of stationary gas must be equal to the weight of that chunk of gas.

11. In what way does the density of an object affect its buoyancy?
12. What two major factors influence the sinking rate of an object?
13. Explain the concept of terminal speed.

9.3

14. What is meant by the term atmosphere?
15. Explain the origin of the earth's atmosphere.
16. What factors influenced the development of the atmosphere to its present form?
17. Why is there so little hydrogen and helium in the earth's atmosphere?
18. Show how the atmosphere acts as a "carrier" in the CO_2 and O_2 cycles.
19. Explain why one would expect the pressure in the atmosphere to decrease with increasing altitude.
20. What two factors cause the observed temperature variation with altitude?
21. What is the ionosphere, and how does it affect radio broadcasting?
22. What is ozone?
23. Explain why hot air rises, in terms of our model for a gas.
24. What sort of air circulation might one expect on a global basis?
25. What are the main factors influencing the movement of air masses?
26. Describe a low and a high.
27. Explain relative humidity.
28. How are clouds formed?
29. Describe the various forms of water vapor in the atmosphere.
30. What is precipitation?

9.4

31. What are the five primary air pollutants?
32. What is the largest source of CO?
33. Why is CO so toxic?
34. How is CO reduction being pursued?
35. Why are there so many possible hydrocarbons?
36. Where do most hydrocarbons come from?
37. What is the major source of SO_x?
38. What is being done to reduce SO_x emissions?

39. Why don't nitrogen and oxygen normally present in the air combine to form NO_x?
40. How are photochemical oxidants formed?
41. Why don't particulates simply settle out?
42. What is the chief device being used to remove visible particulates from smoke?
43. Explain the greenhouse effect.
44. Describe an inversion layer.

Quotes to Ponder

1. "The task of cleaning up the nation's smokestacks is being attacked on so many fronts that ultimate victory seems assured.

At least six methods are being used to attack the problem of air pollution control from smokestacks of power plants and other large industries in the United States, and considerable controversy has developed over their relative merits. According to Alvin Skopp, project manager for Esso Research & Engineering Corp.'s pressurized fluidized combustion program, all the methods ultimately may find wise use.

The simplest way to cut air pollution from a furnace is low sulfur natural gas or oil, but for the big user that's often impossible and in any case expensive. Three other methods involve treating coal and converting it either into low sulfur, low ash synthetic gas, synthetic oil, or a solvent refined coal.

The fifth method is fluid bed combustion either at atmospheric pressure or high air pressure. This is the method that produces the most controversy.

The sixth method of attack is the flue scrubber. Although this admittedly is expensive, Esso's Skopp said the scrubber will continue to be attractive to big users because it is highly effective and it doesn't involve replacing expensive existing boilers as the fluid bed processes would." (Leroy Pope, "Battle of Smokestacks Meets U.S. Ingenuity," United Press Release, October 1, 1972)

2. "All environmental pollution is often lumped in the public's mind, yet the problems are distinct. Water, by transmitting typhoid fever, cholera, and other epidemic diseases, was once a more efficient killer than we have yet managed to make of our air. But sanitary engineers learned to purify water after it had become contaminated. One of man's most dubious contemporary achievements is the ingenuity with which he first pollutes, then purifies, his drinking water.

But once foul air is abroad, man cannot cleanse it. In times of crisis he can turn only to gas masks. Scientists are suggesting today that if we continue to pollute the atmosphere at our present rate, the process may soon become irreversible. We will have destroyed 'this most excellent canopy, the air.'" (Frank Graham, Jr., "The Breath of Death," *Audubon Magazine,* July–August, 1968)

3. "In the field of climate, the sun's radiations, the earth's emissions, the universal influence of the oceans, and the impact of the ice are unquestionably vast and beyond any direct influence on the part of man. But the balance between incoming and outgoing radiation, the interplay of forces which preserves the average global level of temperature, appear to be so even, so precise, that only the slightest shift in the energy balance could disrupt the whole system. It takes only the smallest movement at its fulcrum to swing a seesaw out of the horizontal. It may require only a very small percentage of change in the planet's balance of energy to modify average temperatures by 2°C. Downward, this is another ice age; upward, a return to an ice-free age. In either case, the effects are global and catastrophic." (Barbara Ward and René Dubos, *Only One Earth,* W. W. Norton and Company, New York, 1972)

CHAPTER TEN

A FEW LOOSE ENDS

The first nine chapters of this book have been devoted to understanding solids, liquids, and gases and their role in our environment. We have explored aspects of the traditional disciplines of physics, chemistry, astronomy, and geology along the way, but each was studied only to the depth necessary to get on with our overall goal: a look at our solid, liquid, and gaseous environment from the molecular viewpoint. Now that you've seen physical science at work, perhaps you'll find it interesting to see how physical science is organized in an overall sense, have a brief look at some of the portions of physical science we've missed, and look at some other applications of physical science to modern topics. The future seems to be coming upon us at an ever-increasing pace, so another section of this chapter will be devoted to some projections of what will happen in pure and applied science in the next few decades. Last, but certainly not least, we'll discuss the role of physical science in the overall environmental action picture and make some suggestions about *your* role.

10.1

The Rest of Physical Science, and Some Applications

The parts of physical science we've seen so far have been only those portions we've needed to analyze solids, liquids, and gases. There are a number of other branches, many of which can be applied to parts of our modern environment. Although a hard and fast division of the physical sciences into various topics is just not possible because of the dynamic nature of science and the reality it is grappling with, Figure 10.1 should give you some crude idea of where some things fit into the overall scheme.

Let's examine each of these branches to see what they study and how they go about it.

Physics: The study of matter, motion, and energy in most general terms.

Chemistry: The study of matter and the interactions between different kinds and states of matter.

SHARP STUDENT: Could we stop right here for just a minute, please? First you separated all of physical science into nice, neat little categories, and next you say that both physics and chemistry include the study of matter. Your division must not be too good, because it sure looks like there is some overlap here.

OLD PROFESSOR: I'm afraid you've got me there. There is definitely an overlap between chemistry and physics. In fact, I've known people who called themselves chemical physicists and others who called themselves physical chemists, and yet they did roughly the same type of work.

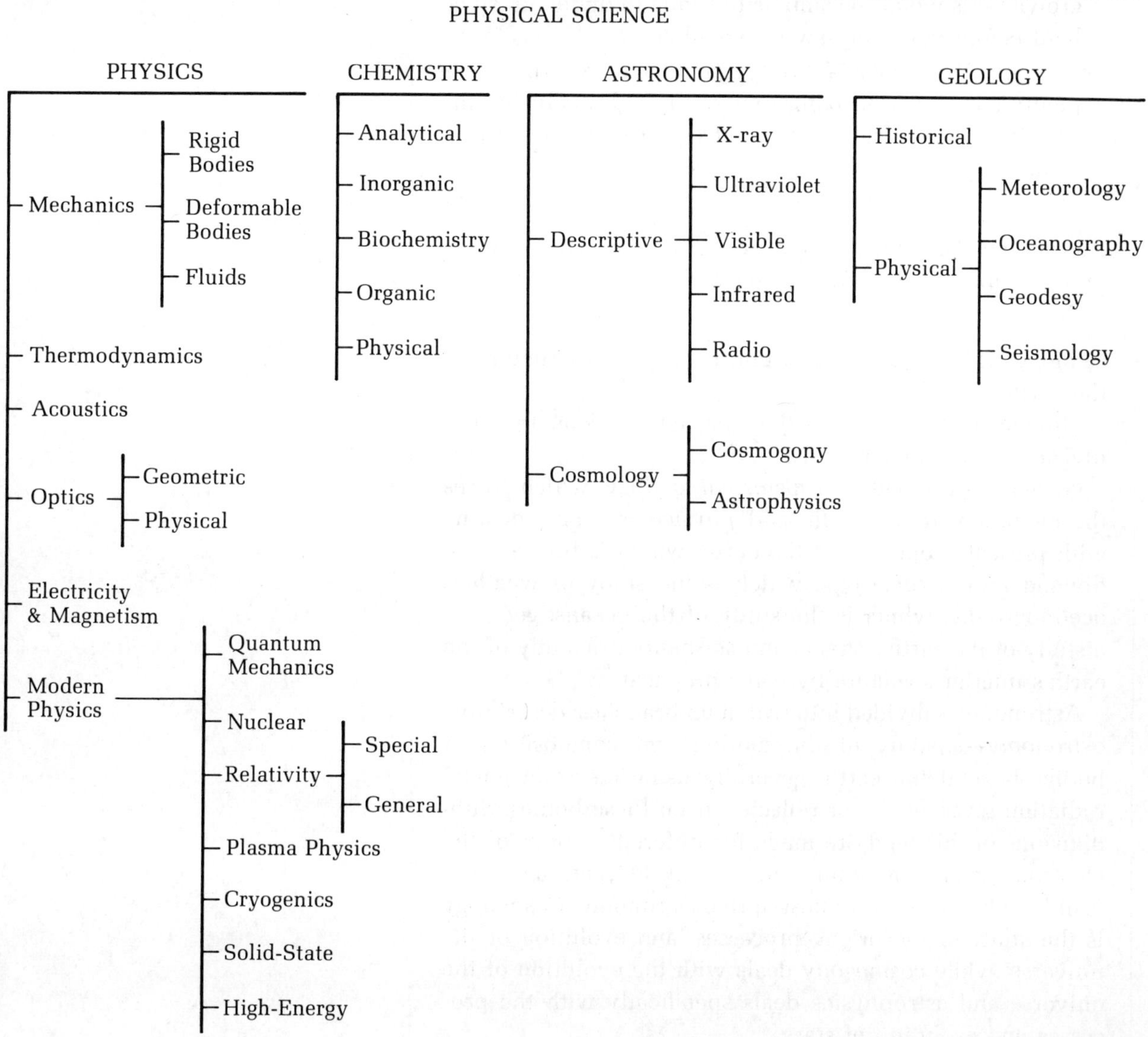

Figure 10.1 *Physical science organization chart*

The division between chemistry and physics is largely historical, and based at least partially on the early alchemists' work of changing one element into another (preferably gold), while the early physicist was busily analyzing the motion of cannonballs.

In modern times, physics and chemistry have grown toward each other and have become hopelessly intertwined in a number of areas.

STUDENT: Oh, I see. These divisions aren't supposed to be too sharp, because they were made some time ago, and the fields have expanded as people's understanding has grown. This would be similar to trying to divide up some land before you ever saw it, wouldn't it?

PROFESSOR: Yes, that's a pretty good analogy. You'll see plenty more overlap before we're through analyzing the structure of physical science. Let's go back to our discussion:

Astronomy: The study of the origin, motion, composition, and evolution of the universe beyond the earth.

Geology: The study of the origin, history, and structure of the earth.

Moving from the last to the first, let's look at the subdivisions of the major fields:

Geology is divided into *historical geology,* which traces the evolution of the earth, and *physical geology,* dealing with present properties of the earth, which is further subdivided into *meteorology,* which is the study of weather; *oceanography,* which is the study of the oceans; *geodesy,* a study of the earth's shape; and *seismology,* a study of the earth's interior structure by wave propagation.

Astronomy is divided into two main branches: *descriptive astronomy*—a study of the motion and composition of bodies beyond the earth, generally using electromagnetic radiation given off by or reflected from these bodies. Subdivisions of this field are made for different regions of the electromagnetic spectrum. Until about 40 years ago, only visible light was used in descriptive astronomy. *Cosmology* is the study of the origin, processes, and evolution of the universe, while *cosmogony* deals with the evolution of the universe and *astrophysics* deals specifically with the processes and evolution of stars.

Chemistry is divided into the following categories: *analytical chemistry,* which primarily involves identification of unknowns; *biochemistry,* the chemistry of living things; *organic chemistry,* the study of compounds which contain carbon; and *inorganic chemistry,* a study of all compounds *not* containing carbon; and *physical chemistry,* the study of the energies involved in interactions between different kinds of matter.

Physics is traditionally divided into *mechanics,* a study of the motion of bodies [and further subdivisions of mechanics are as follows: *mechanics of rigid bodies,* a study of the motion of bodies which retain a fixed shape; and *mechanics of deformable bodies,* a study of the motion of bodies in which the shape of the body changes as the body moves; *fluid mechanics,* a study of the motion of liquids and gases]; *thermodynamics,* a study of the relations between heat and other forms of energy; *acoustics,* a study of sound; *optics,* a study of electromagnetic radiation, including visible light [further subdivisions of optics are *geometric optics,* a study of radiation using the particle theory of light ($\lambda \ll$ apparatus size); and *physical optics,* a study of radiation using the wave theory of light ($\lambda \sim$ apparatus size)]; *electricity* and *magnetism,* the study of all phenomena relating to electric charges and their motion; *modern physics,* a collective term for a group of studies which have arisen since roughly 1900 and which include *quantum mechanics,* a study of the structure of atoms and molecules using the notion of the smallest, indivisible unit of energy; *nuclear physics,* a study of the structure of the nucleus; *relativity,* subdivided into *special relativity,* a modification of old ideas of time and space when a body is moving at constant velocity approaching the speed of light; *general relativity,* modification of old ideas of gravitational force acting on an accelerating body to the thought of space being curved locally in the vicinity of a massive body; *plasma physics,* a study of the fourth state of matter, a plasma, consisting of a gas-like collection of charged particles; *cryogenics,* a study of the behavior of materials at very low temperatures; *solid state,* a study of the electrical properties of semiconducting materials; *high-energy physics,* a study of the fundamental particles and their properties which result from high-energy interactions between other particles.

PATIENT STUDENT: That's quite a list. Is it true that all scientists use the "scientific method" that we've seen so often throughout this book?

PROFESSOR: Not exactly. What they do is generally what fits best their temperament and training. Some scientists are excellent *observers* who fill notebooks, photograph albums, reels of audio tape, etc., with careful, methodical descriptions of phenomena. Others fill pages and pages with ideas about the underlying laws which govern phenomena. Their *hypotheses* may range from the very simple to the most abstract mathematical formulations. Still other scientists delight in taking the hypotheses of others and extending them in new directions, applying them to new situations and *predicting* the results of as-yet-undone experiments. Finally, there are those who work in the laboratory, designing and building equipment, getting the "bugs" out of it, reducing any inaccuracies to the lowest possible level, and carefully, painstakingly, running *experiments* to either confirm or deny predictions made on the basis of some hypothesis. Some scientists work in more than one area, but scientists usually concentrate on one or two parts of this overall sequence.

EAGER STUDENT: To get this down to a more personal level, could you tell us what sort of activities a particular kind of scientist might do in his normal work?

PROFESSOR: That might be difficult, because a scientist might be engaged in any of the fields listed above, working in any of the parts of the scientific method, and he could be employed by a college, by a private corporation, or by the government. Let's look at the kind of work done by a high-energy physicist at a college and an organic chemist working for a private corporation.

Our average high-energy physicist (the particles he works with have high energies, but he doesn't necessarily have high energy) is a combination predictor/experimenter. Let's say he has applied some of the very latest theories to the collision of two particular particles and has designed an experiment to test the predictions. The data he gets to analyze will be pictures of tracks in a bubble chamber made by the high-energy particles speeded up by an apparatus called an accelerator as they come crashing through the chamber. Just as a high-flying jet leaves a trail even though it is so far up as to be almost invisible, high-energy particles leave a trail of bubbles in liquid hydrogen at a temperature above its boiling point. Bubbles of the gas form in the liquid on the ions created by the high-energy particle zooming through the liquid. Photographs of these tracks are taken (see Figure 10.2) with stereo cameras so that all three dimensions are covered.

The photographs must be analyzed to determine which particles are present, their charges, masses, and velocities. The whole apparatus is in a magnetic field, so charged particles "bend" one way or the other, depending on their charge. The output of the bubble-chamber experiments is a series of photographs. Since the exact event desired by the experimenter cannot be forecast precisely, photographs are simply taken at regular intervals and presented to the experimenter, who sorts through them until he finds photographic records of the event he's looking for. At one

Figure 10.2 *Photograph taken in a 30-inch bubble chamber*

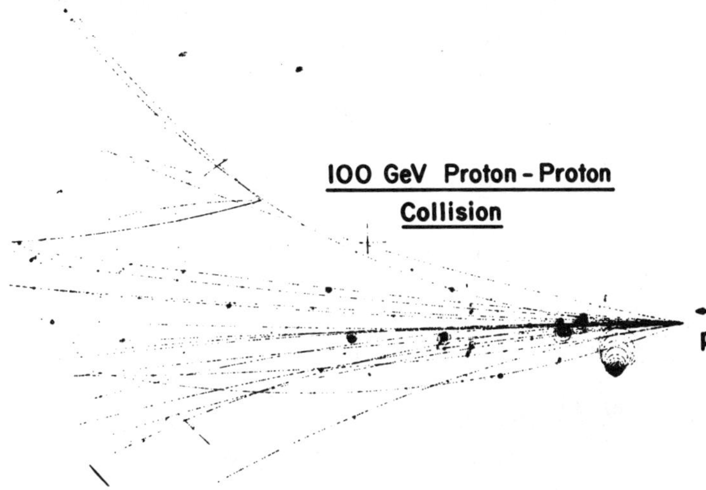

time, this sorting was all done by hand, and one particular person spent several months looking through 200,000 bubble-chamber photographs to obtain about a dozen shots of the event he was looking for. This initial sorting process is now often done by a computer.

SHARP STUDENT: Aha! I wondered if you were going to get around to mentioning computers. They really aren't a part of science directly, are they?

PROFESSOR: No, computers are not a part of science although scientists participated in their development and use them often. After we finish these biographies, we'll discuss computers and their development in showing the difference between science and technology.

Once the pictures from the bubble chamber have been processed by the computer, which has been programmed to scan the pictures for particular patterns, photographs selected as promising are analyzed in detail, and the experimenter decides which, if any, of the predictions are correct. It's a lot quicker describing it than doing it. First, the accelerator, bubble chamber, and camera system must be designed, built, checked out, etc. The physicist must understand the current theories with all their possible mathematical complexities before he can make any sensible predictions, then he must program the computer to "recognize" the particular pattern he has predicted.

In addition to this basic research outlined above, the physicist may have part- or full-time teaching duties with the preparations that go along with teaching (often more fruitful for the professor than for the students); and he may direct the research activities of several graduate students. Further, the physicist must read and study articles in the professional journals relevant to his field of specialization and prepare articles describing his own research. Additionally, the physicist may also participate in the governance of the college by serving on various committees or boards.

As another example, let's look at an organic chemist working for a private corporation, like a drug firm. His problem might well be the synthesis of some "new" organic compound. After checking published literature and talking to colleagues about any unsuccessful procedures attempted in the past, he might modify an old procedure or try a new one with systematic variations of pressure, temperature, concentration, catalysts, time, etc., and obtain several products, any or all of which might be the "new" organic compound he seeks to obtain. He would then analyze the products both qualitatively and quantitatively to find out whether he did make what he set out to make. If he was unsuccessful, he might go back and modify the procedure or even start over with a new procedure. If he was successful, he might measure some of the properties of the new compound and report its synthesis and properties in one of the chemical journals.

Just as in the case of the physicist, it's a lot faster to describe the process than to do it. Searching the literature and colleagues' minds may be a slow process, to say nothing of designing an experimental apparatus (heaters, coolers, boilers, evaporators, distillers, mixers, pressure chambers, which must be the proper size, material, and shape, and may be ordered from different suppliers, take varying amounts of time to arrive, and come in damaged, broken, or just plain wrong). Even after the apparatus is set up and running with all the "bugs" out, the products must be analyzed qualitatively and quantitatively, which is no mean feat. In addition to the duties described above, the chemist may have some managerial duties with the company and may supervise the activities of junior chemists, technicians, or typists.

STUDENT: I certainly must admit they have a lot to do. But somehow they don't seem to do anything that applies directly to people.

PLEASED PROFESSOR: Then you got the point. What scientists are trying to do is to understand physical phenomena by discovering and refining our understanding of the universal laws which control and govern the behavior of things. It is the job of the *engineer* or *technologist* to take the phenomena or the universal laws and use them to design devices or goods that are useful to humans.

Since you asked about computers earlier, let's look at the development of computers as an example of the science-technology interaction. The idea of having a machine perform arithmetic operations is an old one, indeed. Yet computers were not developed by technologists until a much more basic understanding of the electrical properties of materials was obtained by scientists. At first, computers used vacuum tubes, but as scientists learned more about semiconductors in the field of solid-state physics, this knowledge was applied directly to computers, resulting in a tremendous increase in speed and capacity of computers. This is a bit oversimplified, because scientists and technologists often work together and are sometimes even the same people. Yet, I hope you see the difference, which lies in either seeking to understand natural phenomena or using the understanding for some purpose.

STUDENT: I think I see the difference, but you really didn't explain what computers are all about.

PROFESSOR: Quite true, I didn't. There are a substantial number of topics in modern life that have physical science implications in their roots. There really isn't enough room to discuss much more in detail, so I thought I'd give you a list of topics with a few brief explanatory words and let you explore on your own.

STUDENT: Thanks, it's just what I always wanted.

PROFESSOR: Well, here goes a list of topics (in no particular order):

Noise Pollution: Longitudinal waves of sound from many sources constantly impinge on our ears. Look up noise suppression, acceptable levels, wave propagation in moving or heated air, etc.

Sonar: Underwater sound waves reflected by objects bounce back to the source. Given the wave speed and the time, the distance may be calculated.

Crushable Auto Bodies and Bumpers: Kinetic energy of

collision is absorbed by deflection of material and converted to random kinetic energy.

Lasers: Light Amplification by Stimulated Emission of Radiation. Atoms are "pumped" to a higher energy level by shining radiation of just the right wavelength on them; when they jump back to their original energy level, the radiation each atom gives off stimulates emission from another atom, all in phase, adding crests to crests and troughs to troughs.

Holograms: Special pictures taken with laser light which give three-dimensional effects.

Superconductivity: At low temperatures, some metals exhibit no resistance to an electric current. Extremely powerful electromagnets have been made using superconducting coils.

Airplanes and SST: Fluid moving over the top of an airfoil creates a lower pressure than on the bottom, resulting in lift. The SST may fly high enough to leave some of its hydrocarbon, CO and NO_x, emissions in regions of the atmosphere usually occupied mostly by ozone.

The Space Program: Rockets work by expelling mass at high velocity. (Ever try to move a canoe by throwing a paddle? Don't.) They have to get enough kinetic energy to escape the gravitational potential energy well of the earth.

Drugs and Food Additives: Particular molecules fit into the body's complex chemistry in strange ways. Some molecules interrupt pain messages, others generate other messages; some food additives retard spoilage, others are only flavoring and coloring.

Quasars, Pulsars, and Neutron Stars: The nonoptical branches of astronomy have come up with a number of new sources of radiation in the universe.

STUDENT: I guess looking those up might keep me busy for a while.

PROFESSOR: Knowing as much physical science as you should by now, you should find many of these topics quite understandable.

10.2

The Future of Physical Science—Breakthroughs Are Difficult to Schedule

Predicting the future has always been a risky business. Some prominent scientists from the middle 1800's made the confident forecast that all important physical phenomena had been discovered, and that the only task left for future scientists would be to improve the accuracy of the measurements. Forecasts of the future have come a long way since that grand *faux pas*. The difficulty now is not one of determining whether or not change will occur, but which of many possible changes will occur, and in what order. Some futurists argue that modern man is being confronted with an ever-increasing number of changes and thus an ever-increasing number of possibilities for the future. They claim that it is necessary to use the past as an indicator and make short- and long-range forecasts of the future, revising these forecasts as time progresses. If this is not done, the future will "sneak up" on you and you'll be confronted with a situation in which you must make a rapid choice without being aware of all your alternatives.

Figure 10.3 *W. H. Ump making a breakthrough*

Because of the sequential nature of the scientific method, it is possible to make some forecasts about science. The scientific method proceeds by making hypotheses about general laws underlying observations. These hypotheses are used to predict the results of potential experiments; and the experiments are performed to see if the predictions and hence the hypotheses are confirmed or contradicted. As hypotheses are partially contradicted, they are modified and improved until they reflect reality more faithfully. If this sequence is not complete with respect to a particular topic, some scientists will be working diligently to close the loop. As a future forecast, then, one can list unexplained observations and predict that they will become explained by some satisfactory hypotheses. For example, small variations observed in the orbit of the planet Mercury were explained by Einstein's general theory of relativity. Currently, astronomers have observed quasi-stellar objects that they have called *quasars,* and have noticed other objects with rapidly varying radiation outputs called *pulsars.* No completely satisfactory hypothesis has been advanced to explain these phenomena yet.

Another set of phenomena which lack a satisfactory explanation come under the general heading of "fundamental particles." (See Figure 10.4.) High-energy collisions have yielded well over 100 particles besides the usual protons,

Figure 10.4 *Some fundamental particles*

neutrons, and electrons. According to various classification schemes, there are *neutrons, mesons, muons, kaons, hyperons,* and others. Many suspect that the situation is analogous to the development of the periodic table for atoms. Some hypotheses have been proposed, but none have been completely verified.

A second kind of "safe" prediction for the future is that sketchy or incomplete hypotheses will become more complete. The theory of the nuclear force between protons and neutrons is not completely developed, and the weak interaction which is involved in radioactive decay processes is far from being completely understood. "Suspicious" theories will be thoroughly investigated and may well be modified. Some scientists are quite unhappy about the uncertainty principle, which is an unavoidable consequence of the theory of quantum mechanics. Although quantum mechanics has stood up quite well, with its predictions being nicely verified by experiment so far, all it will take will be one clear experimental violation to send the theorists scurrying back to their drawing boards. Einstein's general theory of relativity is also held suspect by some. Because of the nature of the theory, it is difficult to make predictions which can be checked accurately by experiment. Nevertheless, experiments will be performed to attempt to verify this theory. Competing theories will also be decided as more data is gathered. For example, the "Big Bang" vs. the "oscillating universe" controversy will be resolved by newer and better quality astronomical data.

A third safe prediction for the future of science is that predictions which have been made will either be verified or refuted. The existence of *quarks* has been predicted as a way out of the too-many-elementary-particles quandary. Quarks are thought to be constituents of current fundamental particles, in the same sense that protons, neutrons, and electrons are constituents of atoms, which were once thought to be fundamental. Magnetic poles have always been observed in pairs (north and south), but it has been suggested that a natural *magnetic monopole* exists. *Antimatter* has been discovered experimentally and consists of fundamental particles with charges opposite their normal charges (e.g., an antielectron is an electron with a positive charge), and sometimes other properties are also reversed. It has been theorized that antiatoms, antisolids, antiliquids, antigases, antistars, and antigalaxies exist, and it has even been predicted that there is an antiuniverse composed entirely of antimatter. Although the special theory of relativity predicts that particles do not accelerate to a speed greater than the speed of light, a particle could have always been going faster than light and not be able to slow down to a speed slower than light. Such a particle is called a *tachyon,* and a number of experimenters are searching for tachyons. *"Black holes,"* mentioned in Chapter 6, are being sought experimentally to see if such a strongly gravitationally collapsed state of matter really exists.

The last class of predictions which may be made is that experiments will continue to be performed in almost every area of physical science to continually check the established principles. Since the scientific method sequence is founded on inductive reasoning, the laws are at best tentative and must be continually checked to see if what was true in the past continues to be true in the present, so that we can make the big leap and presume that it will be true in the future.

The future of technology is not nearly as easy to forecast, because the choices made in the technological realm have a very large cultural component. Should we, as a society, deploy our technological resources to build supersonic transports to whisk people around the world in a short time, or should we try to develop a reliable mechanical heart? Should we work on containing the fusion reaction to generate energy without depleting fossil fuels, or should we build a laser antiballistic missile system? Should we synthesize new mind-expanding drugs, or should we develop antischizophrenia chemical treatments? Should we develop new riot-control weapons to be used by police against unruly crowds, or should we work on new plastics which degrade when exposed to sunlight? The basic science necessary to achieve any of these goals is just around the corner. However, the decision about which of these choices to pursue involves a lot more than just physical science.

10.3

The Future of the Environment

SHARP STUDENT: As long as we're talking about the future, and we've got a handle on what's going on with our environment at the molecular level, why don't we figure out what to *do* about our environment in the future?

PROFESSOR: We'll try. First, let's discuss the reasons why specific strategies for action are difficult to evolve; then we'll set some action goals.

As we have seen from just our study of solids, liquids, and gases, our environment is an extremely complex intermeshing of many different cycles of activity. We interrupt or modify these cycles only at great risk, because the effects of our modification may be felt far away from us in time and space. For example, the supersonic transport wouldn't do *just* the one thing it would be designed for, that is to whisk people or things from one place to another. It would also introduce unburned hydrocarbons to a level of the atmosphere where they might interfere with natural ozone production and hence our natural buffer from much of the sun's ultraviolet radiation. In this respect, we must be extremely careful of simplistic solutions to complex problems. Should we ban *all* pesticides? If so, food production will undoubtedly suffer, prices will soar, and some of the poor will starve. You can't do just *one* thing, as P. L. Ump discovers in Figure 10.5.

10.4

The Environmental Problem Has Many More Dimensions Than Just the Physical

Besides the physical aspects of the environment which we have studied, there are biological and social dimensions to the environment and hence to the solution of any environmental problem, as G. R. Ump illustrates in Figure 10.6. The study of natural cycles and processes involving man, animals, and plants all set in their natural environment is well beyond the scope of this book, but it should be undertaken by you to add necessary depth to your understanding of the environment.

The third aspect of the environmental problem is perhaps the most complex one of all—the social. Let's look at some of its parts:

Figure 10.5 *You can't do just one thing, demonstrated by P. L. Ump*

Ethics: What is the relationship between man and his environment in the predominant ethical or moral systems of today?

History: How has man's impact on his environment progressed through time?

Psychology: What motivates people to change their behavior patterns to ones less harmful to the environment?

Law: What legal remedies are presently available for environmental action?

Political Science: What are the political realities of various proposed government controls?

Economics: How can hidden costs of environmental pollution be brought to light?

Sociology: How does man's social interaction influence his relationship with his environment?

These sketchy remarks and questions should only serve to underline the need for you to study all aspects of man.

UNHAPPY STUDENT: This looks like a cop-out. It reminds me of the old Pogo saying: "We have met the enemy, and they are us." It seems like you're saying that the problems of the environment are too complex and interrelated for us to do *anything*.

PROFESSOR: That's not true at all, because doing nothing is an action strategy in itself and a very dangerous one at that. We've gotten into some pretty nasty jams by doing nothing. No, by way of action strategies, I'm going to issue to you a two-pronged challenge:

1. Assess your own personal environmental impact.

Analyze your actions and those of your family, at home and where you work. As Booker T. Washington said, "Put your bucket down where you are." Look at your community's solid-waste disposal system, its water and sewage treatment plants. See what recycling activities are going on in your area. Read, study, and join a discussion group. Analyze the relationship between your work and the environment. Find out your company's policy about pollution. Above all, analyze *thoroughly*. Trace your way through whatever cycles are involved. Once you have gathered all the facts, suggest some alternative procedures and trace their likely consequences.

2. Make sure you and your elected representatives vote intelligently on environmental issues.

I'm not going to try to scare you by saying that we only have x more years before certain doom, but I will predict

Figure 10.6 *G. R. Ump tackles the environmental problem*

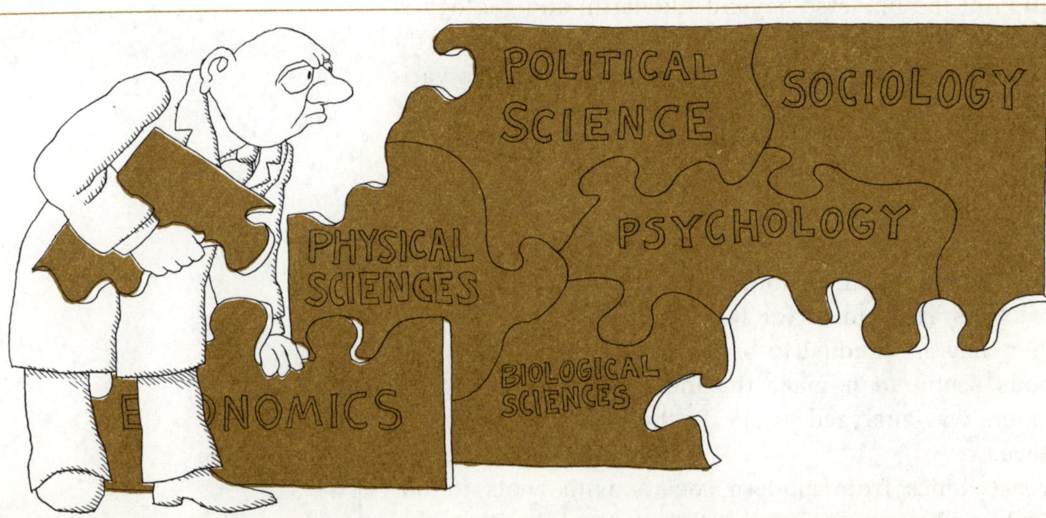

that some very important environmental policy decisions will be made in the next few years. Read what you can about environmental issues (and for that matter, the environmental impact of technological issues), and demand careful, thorough analyses before you make up your mind.

STUDENT: It sure sounds like I'm going to be busy making environmental judgements forever.

PROFESSOR: Welcome to the club! The whole point of this book was to fill in some of the scientific background for you nonscientists so that you won't be awed or buffaloed when the words and numbers start flying in any environmental discussion. You are now well-enough equipped to demand answers from physical science experts about what's happening to the molecules in the environment. After all, you're going to be experts in the social rather than the physical sciences; and it's going to take a balanced approach if we expect to pass along a livable environment to generations to come.

Summary

The physical sciences may be crudely divided into physics, chemistry, astronomy, and geology.

Physics is the study of matter, motion, and energy in most general terms; chemistry is the study of matter and the interaction between different kinds and states of matter; astronomy is the study of the origin, motion, composition, and evolution of the universe beyond the earth; and geology is the study of the origin, history, and structure of the earth. Each of these major divisions of physical science has various subdivisions. Scientists usually specialize not only in some area of physical science, but also in one aspect of the scientific method. As examples, the activities of a high-energy physicist and an organic chemist were analyzed.

The difference between a scientist and an engineer or technologist hinges on whether the physical phenomena and general laws are studied for their own sake (scientist), or whether they are studied to be used for designing devices or goods useful to humans (technologist). The design of computers was analyzed to clarify the science/technology interface.

Several topics from modern society with roots in the physical sciences were mentioned with the hope that you will analyze some of them on your own.

The future is difficult to predict but necessary to ponder, or else the changes will occur so fast that we will be overwhelmed. Because of the sequential nature of the scientific method, some classes of predictions may be made about the future of physical science: unexplained observations will be explained; incomplete hypotheses will be completed; predictions already made will be verified or refuted; and experiments will continue to be performed. The future of technology is not nearly as easy to predict because value decisions must be made before technology can be directed.

Action strategies for us to pursue are difficult to set down because of two great stumbling blocks: you can't do just one thing—simplistic solutions are usually doomed to failure because of the great complexity of environmental processes and cycles; secondly, the environmental problem has many more dimensions than just the physical. Biological implications must be studied in some detail, and social implications are so vast that only the merest hint may be dropped here about what to do.

About all we can see clearly to do is assess our own environmental impact and make sure we and our elected representatives vote intelligently on environmental issues.

Our actions in this regard will determine whether or not we pass along a livable environment for generations to come.

Questions

This may be used like a programmed text by referring to the indicated questions while reading the text.

10.1

1. Name the four major divisions of physical science.
2. Explain the reasons why some of the subdivisions of science overlap.
3. What is the difference between astronomy and geology?
4. What is the difference between organic chemistry and inorganic chemistry?
5. What is the difference between geometric and physical optics?
6. What is the difference between special and general relativity?

7. Outline the kinds of duties performed by a high-energy physicist.
8. How do the duties of a high-energy physicist fit into the "scientific method?"
9. Outline the duties of an organic chemist.
10. What is the difference between a scientist, an engineer, and a technologist?

10.2

11. Why do some claim that the future should be anticipated as clearly as possible?
12. How is it possible to use the "scientific method" to help forecast the future of physical science?
13. List at least one prediction in the four "safe" prediction categories.
14. Why is it more difficult to forecast the future of technology?

10.3

15. Why are there no specific action strategies evolved in this book?
16. What is the meaning of the statement, "You can't do *just* one thing?"

10.4

17. What other dimensions besides the physical enter environmental problems?
18. What two-pronged challenge did the professor issue to the student?
19. Why are nonscience majors who have read this book—you, in other words—uniquely qualified to respond to the challenge of the environment?

Quotes to Ponder

1. "The most persuasive case of ardent environmentalists is that we don't know all the effects of tinkering with the earth. And, of course, they are right. We don't." (H. D. Doan, The Dow Chemical Company, Midland, Michigan)

2. "During the 10-year period 1971–1980, the Council on Environmental Quality estimates that this nation's pollution control cleanup cost will be $287 billion. . . ." (*Environmental Science and Technology*, volume 6, number 9, September, 1972)

3. "As the environment churns faster, we are not only pressured to devote more mental resources to thinking about the future, but to extend our time horizon—to probe further and further ahead. The driver dawdling along an expressway at twenty miles per hour can successfully negotiate a turn into an exit lane, even if the sign indicating the cut-off is very close to the exit. The faster he drives, however, the further back the sign must be placed to give him the time needed to read and react. In quite the same way, the generalized acceleration of life compels us to lengthen our time horizon or risk being overtaken and overwhelmed by events." (Alvin Toffler, *Future Shock,* Random House, New York, 1970)

4. "We may bury our beer cans thereby eventually polluting our ground water. We may burn them, if they are combustible, thereby polluting our air. We may flush them down the drain, if they are soluble, thereby polluting our streams. But we cannot annihilate beer cans. We can destroy the Form but not the Substance of beer cans. Matter is conserved in all chemical transformations, atom for atom." (Henry A. Bent, "Haste Makes Waste," *Chemistry,* October, 1971)

5. "'We can never do one thing,' and we can (in principle) never carry out an absolutely complete analysis, but it is nonetheless possible to institute changes safely. Two intellectual tools make this possible. The first is model-building, as mathematicians understand the term. With an intellectual model, one works out the feedback consequences of alterations in the system in an immaterial realm where the 'costs' can be borne without pain. The second tool is dynamic programing, that is, the use of an adaptive logical system that can correct the model (or its real life analog) rapidly, before run-away feedback processes take control." (Garret Hardin, "To Trouble a Star: The Cost of Intervention in Nature," Bulletin of the Atomic Scientists, January, 1970)

Additional Sources

Here are some books to dig into to explore the nonphysical parts of the environmental problem as well as the physical.

1. Ingraham, Vernon (editor), *Survival-Readings on Environment,* 1971, Holbrook Press, Boston.

2. Day, Fost, and Rose, *Dimensions of the Environmental Crisis,* 1971, John Wiley & Sons, New York.
3. Johnson, Cecil E., *Eco-Crisis,* 1970, John Wiley & Sons, New York.
4. Ward, Barbara and Dubos, Rene, *Only One Earth,* 1972, Norton, New York.
5. Bent, Henry A., "Haste Makes Waste—Pollution and Entropy," *Chemistry,* Volume 44, #9, October, 1971.
6. Henshaw, Paul S., *This Side of Yesterday—Extinction or Utopia,* 1971, John Wiley & Sons, New York.
7. Alvin Toffler, *Future Shock,* 1970, Random House, New York.
8. Schroeer, Dietrich, *Physics and Its Fifth Dimension: Society,* 1972, Addison-Wesley, Reading, Massachusetts.
9. Potter, Van Rensselaer, *Bioethics—Bridge to the Future,* 1971, Prentice-Hall, Englewood Cliffs, New Jersey.

GLOSSARY

A-bomb — Device in which a neutron strikes a large nucleus, causing it to split (fission), releasing energy, more neutrons, and other particles.

Absolute Zero — That temperature at which molecular motion—and, therefore, pressure—ceases.

Acceleration — Rate at which velocity changes; change in velocity divided by time over which change occurs; units: m/sec^2.

Activated Charcoal Treatment — Water treatment which removes some suspended solids by having the solids form bonds with the charcoal (carbon).

Activated Sludge — Water treatment which removes oxygen-demanding wastes by passing the water through sludge containing aerobic bacteria which digest the waste material.

Aerobic Bacteria — Bacteria which require oxygen to complete their normal digestive processes.

Alkali Metals — That group of elements in the vertical column farthest to the left on the Periodic Table: Li, Na, K, Rb, Cs, Fr.

Alpha Rays — One of the three rays given off by radioactive substances; a helium nucleus consisting of 2 protons and 2 neutrons.

Amines — One of the products of the digestive process of anaerobic bacteria, having a "fishy" smell.

Amplitude — The height of a wave above the undisturbed surface.

Anaerobic Bacteria	Bacteria which do not require oxygen to complete their normal digestive process.
Anode	The positively charged plate in a gas discharge tube.
Anti-matter	Matter made of anti-electrons, anti-protons, and anti-neutrons. Anti-electrons have a positive charge; anti-protons have a negative charge. Anti-neutrons differ from neutrons in other ways than charge.
Asteroid	A heavenly body similar to a planet, but much smaller in size.
Astronomy	The science which studies the origin, motion, composition, and evolution of the universe.
Asymmetry	Unevenness; lack of balance.
Atmosphere	The layer of gases called air which surrounds the earth.
Atom	Consists of a nucleus containing protons and neutrons, about 10^{-14} m in diameter, and electrons equal in number to the protons orbiting the nucleus at a distance of about 10^{-10} m with definite energy levels, but only probabilistically known paths.
Beta Ray	One of the three rays given off by radioactive substances; an electron.
Big Bang Theory	A theory of the origin of the universe which says that all matter and energy in the universe was concentrated into a single ball which exploded about 10^{10} years ago.
Biochemical Oxygen Demand (BOD)	The amount of oxygen required to complete the digestion of oxygen-demanding wastes.
Biodegradability	The process by which a substance consisting of large molecules is broken down by natural processes to smaller molecules, which then enter natural cycles.
Black Hole	An object theorized—but not yet observed—to be a star whose mass is so large and gravitational pull so strong that no radiation escapes.

Brownian Motion	The motion of a particle large enough to be seen under a microscope in response to collisions with molecules or groups of molecules in constant, random motion.
Bubble Chamber	Device which tracks the trails of fundamental particles as they shoot through superheated liquid hydrogen, causing bubbles to form.
Cathode	The negatively charged plate in a gas discharge tube.
Cathode Rays	The rays that are given off by the cathode in a gas discharge tube; electrons.
Centi	Metric system prefix meaning one-hundredth; 1/100, 10^{-2}; e.g., a centimeter = 10^{-2} meters.
Chain Reaction	Process by which neutrons given off by one fission cause more fissions, which cause more fissions, etc.
Chemistry	The study of matter and the interactions between the different kinds and states of matter.
Chlorine Treatment	Water treatment which kills bacteria with free chlorine, which takes electrons from bacteria molecules to form chlorine ions.
Cleaving	Process by which some crystals split along smooth, shiny faces.
Colloid	Extremely small (10^{-8} m or less) solid particles which do not dissolve, but sink slowly in a liquid.
Continental Drift	The moving apart of the continents—thought to be caused by the motion of plates in the earth's crust—floating on the mantle.
Continuous	Capable of being subdivided indefinitely.
Cosmology	Study of the origin and evolution of the universe.
Coulomb	Basic unit of electrical charge.
Covalent Bond	Bond which holds atoms together by sharing electrons, forming molecules.

Crest	The upper or peak part of a wave.
Critical Mass	That amount of mass of radioactive material necessary to permit a chain reaction to occur.
Crust	The outermost layer of the earth, consisting of plates 10–50 km thick.
Current, Mantle	Slow movement of the earth's noncrystalline mantle, causing the crustal plates to move and thereby causing earthquakes and volcanoes.
Cycle	Any process which returns to its original state periodically.
DDT	Dichlorodiphenyltrichloroethane; used as an insecticide.
Deductive Reasoning	Reasoning based on deriving certain specific truths, given general truths or other certain specific truths. For example, all authors are weird. Wiggins is an author; therefore, Wiggins is weird.
Density	The ratio of mass to volume for a given substance; units: kg/m^3 or g/cm^3; $D = m/V$.
Detergent	Synthetic organic compound whose molecules have a symmetric end which dissolves other symmetric molecules (like fats, greases, and oils) and an unsymmetric end which dissolves in water.
Diffraction	Phenomenon in which straight waves of light go through an opening of nearly the same size as the light wavelength and come out curved.
Dipole	Any object which has at least two, distinct charge regions (one positive and one negative) whose centers do not coincide.
Direct Conversion	Process by which electromagnetic radiation from the sun is converted directly to electric potential energy.
Discrete	Separate; distinct; incapable of further subdivision.
Dissolved Oxygen (DO)	The amount of gaseous oxygen held in solution by water.
Elastic Collision	One in which the sum of the kinetic energies of the colliding particles is unchanged by the collision.

Electrical Force — That attractive or repulsive force which has been observed to obey Coulomb's law:

$$F_E = K\,\frac{Q_1Q_2}{r^2}.$$

Electromagnetic Radiation — That form of energy which is created when a charge is accelerated, causing a disturbance or wave in the electromagnetic field.

Electromagnetic Spectrum — The whole group of waves generated by accelerated charges, differing in wavelength and frequency, but all traveling at the speed of light.

Electron — A negatively charged fundamental particle whose mass is approximately 1/2000 that of the proton or neutron.

Electrostatic Precipitator — A device which traps solid particles in smoke by their electrical attraction to a charged plate.

Emulsion — Formed when two liquids mix whose molecules do not "fit" together to form a solution; one liquid forms small particles which settle out slowly.

Endothermic Reactions — Reactions which require the addition of kinetic energy to proceed.

Energy — A quantity possessed by a body by virtue of its motion (kinetic energy) or position (potential energy); units: joules = $kg \cdot m^2/sec^2$.

Entropy — A quantity proportional to the disorder in any system.

Escherichia coli — A harmless bacteria found in fecal material.

Eutrophication — A process by which a lake fills up with sediment and decaying organic matter, getting shallower and shallower until it changes from a lake to a swamp to a meadow.

Evaporation — Process by which molecules, with more than average kinetic energy near the surface of a liquid, escape and become part of the surroundings.

Exothermic Reaction — Reaction in which kinetic energy is given off to the surroundings.

Extrapolation Extending a line on a graph beyond the region where experimental data exists.

Fission Process in which a large nucleus is struck by a neutron and splits, resulting in energy, more neutrons, and other particles being given off.

Force A push or a pull; may be caused by gravity, electricity, nuclear force, or the weak force; units: newtons = kg · m/sec^2.

Frequency Number of cycles of a wave in one second.

Friction Transformation of organized kinetic energy of a large object into random kinetic energy of molecules.

Fundamental Particles Particles which cannot be (easily) broken down to other particles; e.g., protons, neutrons, and electrons.

Fusion Process by which protons and neutrons come together to form larger nuclei, releasing energy in the process.

Gamma Ray A ray given off by radioactive substances; a short wavelength electromagnetic radiation.

Geology The study of the origin, history, and structure of the earth.

Geothermal Energy Energy generated by causing the hotter, interior parts of the earth's crust to turn water to steam, driving a turbine.

Glaciers Large ice masses which made several advances and retreats across some of the northern United States.

Gravitational Force That attractive force observed to follow Newton's law of gravitation:

$$F_G = G\frac{m_1 m_2}{r^2}.$$

Greenhouse Effect Atmospheric effect in which the sun's radiation passes through a CO_2 layer; the earth's lower frequency radiation is reflected by the CO_2 back to earth, keeping the earth warm, similar to the operation of a greenhouse.

Halogens Those elements in the second-from-the-right vertical column (next to the noble gases) in the Periodic Table: F, Cl, Br, I, At.

H-bomb — Device in which protons come together to form larger nuclei (fusion), releasing energy and other particles in the process.

Heisenberg Uncertainty Principle — Principle from quantum mechanics which states that the position and momentum of any object cannot be determined exactly at the same time.

Hemoglobin — That part of blood which carries oxygen.

Hologram — Photograph using special laser light that has three-dimensional effects.

Hydrocarbons — Substances whose molecules contain the atoms hydrogen and carbon.

Hyperon — A recently discovered fundamental particle.

Immiscibility — Two liquids whose molecules do not fit together are said to be immiscible.

Incineration — Breaking down large hydrocarbon molecules by combining the carbon and hydrogen with oxygen at high temperatures to form CO_2 and H_2O.

Induced Charge — Charge induced on the surface of a conductor by the presence of another charge which repels like charges and attracts unlike charges, leaving unlike charges closer together than like charges.

Induced Dipole — Molecule which acts as a dipole in the presence of another charge, which may be just another molecule, and forms bonds by the induced charge effect.

Inductive Reasoning — Form of reasoning which proceeds from a certain specific truth to an uncertain generality. For example, I have known three English majors; all three were blondes. Therefore, all English majors are blondes.

Inelastic Collision — One in which kinetic energy of one or both bodies is converted to potential energy as a result of the collision.

Inertia — Resistance to change, especially resistance to the change of velocity of a moving body.

Infrared Ray — Form of electromagnetic radiation with wavelength slightly longer than visible light.

Insolubility A solid whose molecules don't fit a liquid's molecules is said to be insoluble.

Interference Interference occurs when two or more waves attempt to occupy the same point at the same time.

Inversion Layer Atmospheric condition in which a warmer layer of air overlays a cooler layer, contrary to the normal decrease in temperature which allows upward air motion.

Ionization Process by which atoms gain or lose electrons, resulting in filled outer energy levels.

Ionosphere Layer of the atmosphere containing a large proportion of ions and free electrons.

Ions Atoms which have gained or lost enough electrons to result in a filled outer energy level.

Isotopes Different forms of the same element, differing only in the number of neutrons in the nucleus, but not in chemical properties.

Joule The unit of energy; $kg \cdot m^2/sec^2$.

Junk Material which is in the wrong place at the wrong time to be considered useful.

Kaon Recently discovered fundamental particle.

Kilo Metric system prefix meaning one thousand, 1000, 10^3.

Kilogram 10^3 grams; the fundamental unit of mass in the metric (MKS) system.

Kinetic Energy (*KE*) Energy due to motion, $\frac{1}{2}mv^2$.

Laser Acronym for <u>L</u>ight <u>A</u>mplification by <u>S</u>timulated <u>E</u>mission of <u>R</u>adiation. The light is emitted when electrons jump from higher to lower energy levels in atoms.

Longitudinal Waves Waves in which the medium is alternately compressed and expanded in the same direction as the wave moves.

L-waves One of three seismic waves; the slowest moving wave which travels along the earth's surface.

Mantle — That portion of the earth just below the crust, consisting of hot, dense, noncrystalline rock, which moves slowly like a "stiff" liquid.

Mass — Amount of matter, material, or stuff possessed by a body; unit: kg.

Mega — Metric system prefix standing for one million; 1,000,000; 10^6; e.g., a megameter = 10^6 meters.

Meson — A recently discovered fundamental particle.

Mesosphere — A middle atmospheric layer from 50 to 80 km altitude

Meter — The basic unit of length in the metric (MKS) system; 10^7 meters = the distance from pole to equator.

Methane — One of the decay products of anaerobic bacteria; CH_4.

Methyl Mercury — Compound of methane and mercury soluble in water.

Micro — Metric system prefix standing for one-millionth; 1/1,000,000; 10^{-6}; e.g., a microgram = 10^{-6} grams.

Milli — Metric system prefix standing for one-thousandth; 1/1000; 10^{-3}; e.g., a millisecond = 10^{-3} seconds.

Miscibility — Two liquids whose molecules "fit" together are said to be miscible.

MKS System — Acronym for the Meter-Kilogram-Second system of units.

Model — Something which represents a concept different from itself; e.g., a model airplane represents a real airplane.

Molecule — Two or more atoms bonded together by a covalent bond.

Momentum — The product of mass and velocity; unit: kg · m/sec.

Muon — Recently discovered fundamental particle.

Natural Dipole — Molecule which has an unsymmetric charge distribution, and bonds + region to − region with other molecules.

Natural Philosophy — Forerunner of science, practiced mostly on a theoretical basis by people such as Aristotle.

Neutrino — Recently discovered fundamental particle.

Neutron — Fundamental particle of mass equal to that of the proton, almost 2000 times that of the electron, and having no charge.

Newton — Unit of force; kg · m/sec^2.

Newton's First Law — A body in motion in a straight line tends to continue its motion in a straight line, and a body at rest tends to stay at rest.

Newton's Second Law — $F = ma$.

Noble Gases — Those elements in the vertical column farthest to the right in the Periodic Table: He, Ne, Ar, Kr, Xe, Rn.

Node — That part of a wave which is at the same height as the undisturbed surface.

Nova — The explosion of a small star which has manufactured too many heavy nuclei.

Nuclear Reactor — Device for generating energy by fission of large nuclei.

Nucleus — That portion of an atom about 10^{-14} m in diameter containing protons and neutrons.

Open Dump — Accumulation of solid waste in a single location with no treatment.

Organic — Molecules which contain carbon are organic.

Oscillating Universe — Theory which says that the universe originated in a primeval fireball which exploded, but the explosion will show and turn around, eventually reforming the fireball, which will explode, etc.

Ozone — Molecule consisting of three oxygen atoms.

Paraffin — Noncrystalline substance made from hydrocarbon molecules of different lengths.

Particulates — Particles carried along in the air.

Period — Time necessary for an oscillation to complete one full cycle.

Pesticide	Substance whose function is to kill pests, especially insects.
Phosphates	Materials which contain any of several phosphorus-oxygen groupings.
Photochemical Oxidants	Substances which combine with the oxygen in the air in the presence of sunlight.
Photons	Light; considered to be a particle.
Photosynthesis	Process by which plants combine CO_2 and H_2O in the presence of sunlight and other nutrients to form large hydrocarbon molecules.
Physics	The study of matter, motion, and energy in most general terms.
Plate	A portion of the earth's crust.
Polarity	A measure of the unevenness or asymmetry of the charge distribution of a molecule. The more uneven the charge distribution, the more the polarity of the molecule.
Pollution	Air or water is polluted when it contains substances which render it unsuitable for an intended use.
Potential Energy (*PE*)	Energy due to position; may be gravitational (mgh), electrical or nuclear.
Precipitation	Occurs when a dissolved substance comes out of solution, especially a solid out of a liquid, or a liquid out of a gas.
Pressure	The force per unit area; unit: nt/m^2.
Primary Treatment of Sewage	Process of settling to remove some suspended solids, and chlorination to remove bacteria.
Primeval Fireball	According to the Big Bang theory of the origin of the universe, all the matter and energy in the universe was contained in one hot, dense glob called the primeval fireball.
Proportionality	Directly proportional means that two quantities vary in the same way; when one goes up, so does the other go up; when one goes down, so does the other go down.

Proton — Fundamental particle whose mass is almost 2000 times that of the electron, and is positively charged.

Pulsar — A star whose brightness varies widely; i.e., pulsates.

P-wave — A seismic wave which travels fastest and is a longitudinal wave.

Qualitative — Having to do with general characteristics or tendencies.

Quantitative — Having to do with something that can be measured and has some numerical value and unit.

Quantum — The smallest possible unit of electromagnetic radiation.

Quark — Hypothetical particles which are thought to be basic building blocks for fundamental particles.

Quasar — Star which has its major energy output in a non-visible part of the electromagnetic spectrum; radiowaves, usually.

Radar — Acronym for Radio Detection And Ranging; a device which bounces electromagnetic waves off objects and calculates their distance away, knowing the bounce time and speed of light.

Radioactivity — Natural fission of large nuclei, producing smaller nuclei, energy, alpha, beta, and gamma rays, and other particles.

Recycling — Process by which some items are returned to an earlier part of a cyclic process.

Relative Humidity — The amount of moisture in the air relative to the maximum moisture it could hold.

Relativity — Theory based on the constancy of the speed of light, which has profound consequences on the nature of time and space.

Sanitary Landfill — Solid waste-disposal method in which a layer of dirt is spread over each layer of waste.

Satellites — Heavenly bodies which orbit other heavenly bodies because of gravitational attraction.

Saturation — Condition which occurs when all available solvent molecules are busy solvating other molecules so that no new ones may be solvated.

Scientific Method — A sequence of activities sometimes followed as science progresses; observation, hypothesis, prediction, and experimentation.

Second — Basic unit of time in the metric (MKS) system.

Secondary Treatment of Sewage — Reduction of oxygen-demanding wastes by aerobic bacteria.

Sediment — Solid particles which settle out of a water suspension.

Seismic Waves — Waves which are generated naturally by earthquakes or artificially by explosions.

Solubility — A substance is soluble if its molecules "fit" the molecules of the substance in which it is placed.

Solvated Molecules — Those molecules surrounded by solvent molecules in a solution.

Speed — Distance traveled divided by time necessary to travel that distance; unit: m/sec.

Standard Scientific Notation — Expression of any number in the form of a coefficient between 1 and 10 times the base 10 to some power.

Stratosphere — Next to the lowest layer of the atmosphere.

Superconducting — Allowing electrical charges to pass with no resistance.

Supercooling — Occurs when a substance made of complicated molecules drops to a temperature below its normal crystallization temperature.

Supernova — The explosion of a large star which has manufactured too many heavy elements.

Superposition, Principle of — When more than two waves come together at a point, the displacement at that point is found by simply adding the displacements of the waves.

Suspension — Formed from solid particles whose molecules do not "fit" liquid molecules and slowly sink to the bottom.

S-wave — Seismic wave which travels at an intermediate speed in solids, not at all in liquids, and is transverse.

Symmetry — Evenness, particularly of electrical charge distribution.

Tachyon — Hypothetical particle which goes faster than light.

Temperature — A measure of the random kinetic energy of the moving molecules of a substance.

Terminal Speed — The maximum speed attained by an object as it sinks in some medium.

Tertiary Treatment of Sewage — Removal of specific pollutants, like phosphates or nitrates.

Thermal Pollution — Raising the temperature of some body of water by transfer of kinetic energy.

Thermosphere — Highest layer of the earth's atmosphere.

Topography — The physical shape of the earth's surface.

Total Energy — The sum of kinetic and potential energy.

Toxicity — Mechanism by which harm is done to some living being.

Transverse Waves — Waves in which the motion of the medium is perpendicular to the motion of the wave.

Trickling Filter Purification — Secondary water treatment in which water is trickled over stones in which aerobic bacteria grow.

Troposphere — Lowest level of the earth's atmosphere in which weather is generated.

Trough — The lowest or "valley" part of a wave.

Ultraviolet Catastrophe — Explanation of the frequency distribution of energy given off by heated bodies which yielded infinite total energy when the individual energies were added up.

Ultraviolet Radiation — Electromagnetic radiation whose wavelength is shorter than the visible part of the spectrum.

Velocity — Distance traveled in a given direction divided by the amount of time necessary to travel that distance; units: m/sec.

Volume Amount of space occupied by a body; units: m^3, cm^3.

Wavelength Distance from one part of a wave to the next place the same part appears; e.g., distance from crest to crest.

Weak Force The fourth, and most poorly understood, force which exists in nature.

Weight The amount of gravitational force exerted on a body at the surface of the earth; unit: nt.

X-ray Electromagnetic wave whose wavelength is considerably shorter than visible light; e.g., approximately 10^{-10} m.

INDEX

2

al ce
nme A